Fostering Science Teaching and Learning for the Fourth Industrial Revolution and Beyond

Garima Bansal
Australian Council for Educational Research, India

Umesh Ramnarain
University of Johannesburg, South Africa

A volume in the Advances in Educational Technologies and Instructional Design (AETID) Book Series

IGI Global
PUBLISHER of TIMELY KNOWLEDGE

Published in the United States of America by
IGI Global
Information Science Reference (an imprint of IGI Global)
701 E. Chocolate Avenue
Hershey PA, USA 17033
Tel: 717-533-8845
Fax: 717-533-8661
E-mail: cust@igi-global.com
Web site: http://www.igi-global.com

Library of Congress Cataloging-in-Publication Data

Names: Bansal, Garima, 1983- editor. | Ramnarain, Umesh Dewnarain, editor.
Title: Fostering science teaching and learning for the fourth industrial revolution and beyond / edited by: Garima Bansal, Umesh Ramnarain.
Description: Hershey PA : Information Science Reference, [2023] | Includes bibliographical references. | Summary: "This book addresses the following questions: How do we prepare prospective teachers for the 4th Industrial Revolution? How do we create teacher education curricula that will help pre-service teachers to be sufficiently versatile in the rapidly changing world? Which key perspectives, processes, methods and tools have especially promising payoffs in the lives of pre-service teachers? How often will they serve to inform teacher/teaching decision making and action?"-- Provided by publisher.
Identifiers: LCCN 2022055057 (print) | LCCN 2022055058 (ebook) | ISBN 9781668469323 (hardcover) | ISBN 9781668469361 (paperback) | ISBN 9781668469330 (ebook)
Subjects: LCSH: Science teachers--Training of. | Science--Study and teaching. | Education--Effect of technological innovations on. | Industry 4.0.
Classification: LCC Q181.A2 F67 2023 (print) | LCC Q181.A2 (ebook) | DDC 507.1--dc23/eng20230203
LC record available at https://lccn.loc.gov/2022055057
LC ebook record available at https://lccn.loc.gov/2022055058

This book is published in the IGI Global book series Advances in Educational Technologies and Instructional Design (AETID) (ISSN: 2326-8905; eISSN: 2326-8913)

British Cataloguing in Publication Data
A Cataloguing in Publication record for this book is available from the British Library.

Advances in Educational Technologies and Instructional Design (AETID) Book Series

ISSN:2326-8905
EISSN:2326-8913

Editor-in-Chief: Lawrence A. Tomei, Robert Morris University, USA

MISSION

Education has undergone, and continues to undergo, immense changes in the way it is enacted and distributed to both child and adult learners. In modern education, the traditional classroom learning experience has evolved to include technological resources and to provide online classroom opportunities to students of all ages regardless of their geographical locations. From distance education, Massive-Open-Online-Courses (MOOCs), and electronic tablets in the classroom, technology is now an integral part of learning and is also affecting the way educators communicate information to students.

The **Advances in Educational Technologies & Instructional Design (AETID) Book Series** explores new research and theories for facilitating learning and improving educational performance utilizing technological processes and resources. The series examines technologies that can be integrated into K-12 classrooms to improve skills and learning abilities in all subjects including STEM education and language learning. Additionally, it studies the emergence of fully online classrooms for young and adult learners alike, and the communication and accountability challenges that can arise. Trending topics that are covered include adaptive learning, game-based learning, virtual school environments, and social media effects. School administrators, educators, academicians, researchers, and students will find this series to be an excellent resource for the effective design and implementation of learning technologies in their classes.

COVERAGE

- K-12 Educational Technologies
- Educational Telecommunications
- Bring-Your-Own-Device
- Hybrid Learning
- Virtual School Environments
- Web 2.0 and Education
- Classroom Response Systems
- Collaboration Tools
- Instructional design
- Curriculum development

IGI Global is currently accepting manuscripts for publication within this series. To submit a proposal for a volume in this series, please contact our Acquisition Editors at Acquisitions@igi-global.com or visit: http://www.igi-global.com/publish/.

The Advances in Educational Technologies and Instructional Design (AETID) Book Series (ISSN 2326-8905) is published by IGI Global, 701 E. Chocolate Avenue, Hershey, PA 17033-1240, USA, www.igi-global.com. This series is composed of titles available for purchase individually; each title is edited to be contextually exclusive from any other title within the series. For pricing and ordering information please visit http://www.igi-global.com/book-series/advances-educational-technologies-instructional-design/73678. Postmaster: Send all address changes to above address.

Titles in this Series

For a list of additional titles in this series, please visit:
http://www.igi-global.com/book-series/advances-educational-technologies-instructional-design/73678

Handbook of Research on Shifting Paradigms of Disabilities in the Schooling System
Hlabathi Rebecca Maapola-Thobejane (University of South Africa, South Africa) and Mbulaheni Obert Maguvhe (University of South Africa, South Africa)
Information Science Reference • © 2023 • 435pp • H/C (ISBN: 9781668458006) • US $270.00

Multifaceted Analysis of Sustainable Strategies and Tactics in Education
Theresa Dell Neimann (Oregon State University, USA) Lynne L. Hindman (Oregon State University, USA) Elena Shliakhovchuk (The Polytechnic University of Valencia, Spain) Marian Moore (Austin Community College, USA) and Jonathan J. Felix (RMIT University, Vietnam)
Information Science Reference • © 2023 • 335pp • H/C (ISBN: 9781668460351) • US $215.00

Promoting Diversity, Equity, and Inclusion in Language Learning Environments
Karina Becerra-Murillo (Jurupa Unified School District, USA & American College of Education, USA) and Josefina F. Gámez (Jurupa Unified School District, USA)
Information Science Reference • © 2023 • 326pp • H/C (ISBN: 9781668436325) • US $215.00

Cases on Effective Universal Design for Learning Implementation Across Schools
Frederic Fovet (Royal Roads University, Canada)
Information Science Reference • © 2023 • 300pp • H/C (ISBN: 9781668447505) • US $215.00

For an entire list of titles in this series, please visit:
http://www.igi-global.com/book-series/advances-educational-technologies-instructional-design/73678

IGI Global
PUBLISHER of TIMELY KNOWLEDGE

701 East Chocolate Avenue, Hershey, PA 17033, USA
Tel: 717-533-8845 x100 • Fax: 717-533-8661
E-Mail: cust@igi-global.com • www.igi-global.com

Table of Contents

Detailed Table of Contents

Facing the increasingly complex and ever-changing environment of present challenges, citizens in the fourth industrial revolution across the globe will need to develop knowledge-in-use proficiencies, which equips them with the scientific knowledge to make evidence-based informed decisions, support policy changes, and understand future consequences of lack of action. However, knowledge-in-use is a complex construct with multiple different sub-components. It remains challenging to both understand the underlying information processing mechanism and the inner cognitive process of knowledge-in-use. This chapter unfolds the latent process of developing knowledge-in-use by referencing several landmark cognitive models. Building on that, this chapter discusses how a project-based learning environment has the potential to support students' knowledge-in-use development.

This chapter provides an insight into the theoretical perspectives which form the foundation of extended reality (XR) and its emergence in practice as a fundamental part of science curricula. Issues such as the authenticity of learning, the validity and reliability of XR within processes of assessment, and the theoretical underpinnings of pedagogical approaches in health professions pedagogy are illuminated. Also considered are the implications of XR within the context of science-based learning and the delineation of cognitive, affective, and psychomotor domains of learning in relation to patient outcomes at the front line of care in applied scientific practice. The Covid-19 pandemic, which has impacted all science based pedagogic practice since March 2020, is also considered in the context of moves to ensure that science education can continue, albeit via hybrid models of learning as opposed to traditional pedagogical approaches which have remained little altered over the last century.

It has often been observed that what is assessed often becomes what is taught. To enable teachers to teach 21st century skills, it is crucial that school systems upskill their models of designing high-quality competency-based assessments. This chapter reports a competency-based formative assessment used to equip students with the skills required for successful survival in the 21st century. It reports a formative assessment task in which story writing is used to develop graph development skills among STEM students studying in middle grades in an Indian school classroom. The results indicate that competency-based assessments unravel student-held misconceptions which can be used by the teachers to provide appropriate formative feedback.

Given the importance of inquiry in science education, various computational tools and emerging 4IR technologies have been embraced in recent years to improve how inquiry is carried out across science disciplines. However, there is growing concern about how schools can better guide students and teachers in leveraging the technological affordances of the 4IR. In response to this, the authors propose that one pertinent approach to the development and preparation of science teachers for the future of 4IR classrooms is to teach them how to think computationally. Based on a recent literature review, this chapter offers insight into how computational thinking can be used as a pedagogical approach to science teaching in a 4IR classroom.

Today, virtual learning environments (VLEs) are increasingly valuable for mediating teaching and learning in higher education. This chapter reports the affordances of VLEs in remote chemistry learning. The effects of VLEs versus remote lectures on students' achievement scores in chemistry content tests were compared, and the affordances of VLEs for 21st-century chemistry learning were investigated. This research aimed to investigate the impact of VLEs when learning chemistry remotely and answer some questions about their affordances in the fourth industrial revolution (4IR). Fifty pre-service science teachers (also referred to as students) from a South African university participated in the studies. A mixed methods approach with a quasi-experimental design was preferred. Pre-and post-tests were used to evaluate achievement scores for control and experimental groups, while focus group interviews were used for data gathering in reflection. Results showed a positive shift in achievement scores of the experimental group compared to the control group.

In this study, classroom interaction is addressed by using automatic speech recognition (ASR) for automatic analysis of teacher talk. Educational research on classroom interactions and talk is mainly based on manual analysis, which, although quite accurate, is time consuming and expensive. As there are already many ways to harness the power of automatic approaches to analyse the content of talk, the next step is the automatic detection of different types of talk. In this study, the latter includes the automatic detection of teacher questions and their differentiation into closed and open types. The data consists of ASR text outputs of 25 physics lessons on the same topic, each 90 minutes long, from 25 different science teachers. The interplay between human- and machine-led approaches resulted in promising steps taken towards automatization. Future possibilities for automatic analysis of classroom talk and the implications for teaching and learning are discussed.

This exploratory study investigated pre-service science teachers' (PSTs') learning in the context of a teacher education activity involving the use of innovative video technology. Specifically, the study investigated the nature of the PSTs' noticing when viewing different videos of their rehearsals of a core practice: eliciting and working with student thinking and PSTs' perceived usefulness of viewing and reflecting on different types of video footage. PSTs' rehearsals were recorded by multiple cameras at various locations, including a video recorder at the rear of the classroom, camera goggles worn by the PST, and a student actor. Findings reveal that the PSTs showed different observation foci, observation specificity, and analytic approach when analyzing different types of footage. Although the PSTs rated viewing all types of footage as useful in enhancing their learning, they gave more favorable ratings on videos filmed from the back of the classroom than the point-of-view footage. The implications of using videos to support PSTs' learning are discussed.

The education system necessitates preparing today's children for the needs of tomorrow, and more so in our changing society educated by a schooling system that is agile and flexible in preparing students for jobs yet to be invented. Education systems may no longer be required to teach students content, and instead need to focus on giving students skills to curate information and demonstrate their capacity to be critical and creative thinkers, problem solvers, communicators, and collaborators. Moreover, these skills will be predominantly useful for knowledge acquisition and the creation of science subjects. Are we preparing students for this new world of ever-increasing demands for knowledge? Are we creating thinkers and active learners? This chapter determines how we can create learners who can demonstrate these transversal competencies and substantiates our claims based on research findings where authors searched beyond 'knowledge concepts' and sought to create more agile and flexible science learners who are prepared for some of the challenges posed by the 4th Industrial revolution.

In this study the authors shall endeavour to find answers to how the development of scientific competencies can be taught to primary school children and which methods can be adopted in order to make science learning meaningful according to an inquiry-based approach in a primary school setting. In section one, the methodology aims at active and meaningful learning, and at developing critical thinking and collaborative skills. The teaching methodology based on inquiry will be highlighted showing how this approach is stimulating and motivating for students, enabling them to apply their research skills, to build on meaning and to acquire scientific knowledge through practical experience. In section two, a programme followed by a third grade class will be described. The learning of science actively involves the students in its different stages - formulating the research questions, planning, and carrying out the research to answer the question all are stimulating activities. The following is a simple example of the 5E methodology, an approach based on inquiry.

In the current Fourth Industrial Revolution (4IR), or Industry 4.0, learning—and consequentially learners' comprehension—either in a discipline-specific (science, technology, engineering, and mathematics- STEM disciplines) or general decision-making, can be marred by inappropriate use of the cognitive tenets, particularly so in the current 4IR or Industry 4.0. Accordingly, the current study is informed by this challenge and in part anchored upon theoretical disposition of 4IR or Industry 4.0 cognitive principles. The present research provides both theoretical and empirical examinations to improve mathematical discipline cognition as well as general decision making, thus aimed to enhance knowledge-building and subsequent development in discipline-specific.

Preface

Welcome to the edited reference book, "Fostering Science Teaching and Learning for the Fourth Industrial Revolution and Beyond." As editors, we are delighted to present this comprehensive compilation of research and insights in addressing the challenges and opportunities faced by science educators in the rapidly evolving landscape of the Fourth Industrial Revolution (4IR).

The 4IR has ushered in a new era characterized by exponential technological advancements, ranging from artificial intelligence and robotics to the internet of things, 3D printing, nanotechnology, and more. As the pace of change accelerates, the future becomes increasingly volatile, uncertain, complex, and ambiguous—the very essence of VUCA. In this context, the development of 21st-century skills and competencies, such as critical thinking, creativity, communication, and collaboration, has become essential to prepare the next generation of learners for success in this transformative era.

This edited reference book aims to bridge the gap between theory and practice by providing science educators, research scholars, and science teachers with a wealth of research-based insights and evidence-based strategies. By grounding our recommendations in rigorous research, we strive to empower educators to navigate the challenges and leverage the opportunities presented by the 4IR. The research presented within these pages addresses critical questions at the heart of science education in the 4IR. We explore how to effectively prepare prospective teachers to be versatile and adaptable in a rapidly changing world. By delving into the design of teacher education curricula, we seek to equip pre-service teachers with the knowledge, skills, and dispositions needed to thrive in the 4IR and beyond.

The chapters presented in this edited book represent a cross-section of international contexts that include school settings in Australia, India, Hongkong, South Africa, Finland, Italy, the United Kingdom and the United States. Each chapter in this book builds upon a foundation of research, drawing from the expertise of our contributors who are actively engaged in scientific inquiry and research. They have shared their valuable insights, experiences, and evidence-based practices, which are intended to inform decision-making and guide effective action in the field of science education.

By highlighting the most promising perspectives, processes, methods, and tools, we present educators with actionable strategies that can drive positive change in their classrooms.

The research-based approaches explored in this book encompass various aspects of science education, such as lesson designs, assessment strategies, classroom discourse, the use of information and communication technology, addressing misconceptions, leveraging visuals, and promoting diversity and inclusivity in science classrooms. Furthermore, we explore the transformative potential of open educational resources, e-learning, mobile learning, and the emerging frontier of virtual reality in science teaching.

By emphasizing the importance of research, we aim to inspire our audience to embrace a spirit of inquiry and engage in their own research endeavors. We firmly believe that research is a powerful tool that can shed light on the most effective instructional practices, inform policy decisions, and ultimately enhance science teaching and learning for the 4IR and beyond.

We hope that this book ignites a passion for research and empowers educators to embark on their own scholarly journeys. Together, let us harness the power of research to navigate the complexities of the 4IR, and shape a future where science education thrives and prepares learners for the opportunities and challenges of tomorrow. A brief overview of all the chapters included in this book is provided here for readers' reference.

THEORY INTO PRACTICE: SUPPORTING KNOWLEDGE-IN-USE THROUGH PROJECT-BASED LEARNING

In the context of the Fourth Industrial Revolution, where global citizens face complex and rapidly evolving challenges, the development of knowledge-in-use proficiencies becomes imperative. This chapter recognizes the need for individuals to possess scientific knowledge that enables them to make informed decisions, support policy changes, and comprehend the future consequences of inaction. However, understanding the underlying mechanisms and cognitive processes involved in knowledge-in-use remains a challenge. This chapter delves into the intricate nature of knowledge-in-use, exploring its various sub-components and referencing landmark cognitive models to unravel its latent process. By building upon this foundation, the chapter explores how a project-based learning environment holds the potential to facilitate the development of knowledge-in-use among students. Drawing upon theory and research, the chapter discusses the conceptual framework and practical implementation of project-based learning as a means to support students' knowledge-in-use development. By bridging theory and practice, this chapter provides educators with valuable insights and practical strategies for implementing project-based learning to support the development of knowledge-in-use among students. It offers a comprehensive exploration of the potential benefits and challenges associated with this pedagogical approach, empowering educators to create meaningful learning experiences that prepare students for the complexities of the fourth industrial revolution and beyond.

PROGRESSIVE PEDAGOGY FOR EXTENDED REALITY: SHAPING AND FACILITATING THE FOURTH INDUSTRIAL REVOLUTION

This chapter delves into the theoretical perspectives that underpin Extended Reality (XR) and its application as an integral part of science curricula. It explores various issues related to XR, including the authenticity of learning, the validity and reliability of XR in assessment processes, the chapter also examines the implications of XR in science-based learning, particularly in the cognitive, affective, and psychomotor domains, and its impact on patient outcomes in applied scientific practice. The chapter considers the context of the Covid-19 pandemic, which has significantly affected science-based pedagogical practices. It explores how XR has emerged as a solution to ensure the continuity of science education through hybrid models of learning, overcoming the limitations imposed by traditional pedagogical approaches that have seen minimal change over the past century. It highlights the importance of incorporating XR into progressive pedagogical approaches, offering authentic and immersive learning experiences. The chapter also addresses the assessment aspects of XR and its implications for health professions pedagogy.

The chapter explores the intersection of XR with science-based learning, emphasizing its role in enhancing the quality and outcomes of scientific practice. It addresses the challenges and opportunities presented by XR, particularly in relation to patient care and the application of scientific knowledge in real-world

CREATIVE STORY WRITING: COMPETENCY-BASED FORMATIVE ASSESSMENTS IN STEAM CLASSROOM

This chapter delves into the importance of designing high-quality competency-based assessments to effectively assess 21st-century skills. It explores a unique formative assessment approach that utilizes creative story writing as a means to develop graph development skills among STEM students in middle grades. By implementing this competency-based assessment task, the chapter demonstrates how it can uncover student-held misconceptions and provide valuable insights for teachers to deliver appropriate formative feedback. The results presented in this chapter shed light on the effectiveness of this innovative assessment method and its potential for nurturing essential skills in the 21st century.

COMPUTATIONAL THINKING IN SCIENCE TEACHING AND LEARNING

This chapter explores the integration of computational thinking into science teaching and learning, considering the importance of inquiry-based approaches in science education. It acknowledges the increasing adoption of computational tools and emerging technologies from the Fourth Industrial Revolution (4IR) to enhance inquiry practices in science disciplines. However, there is a growing concern regarding the need to guide students and teachers in effectively leveraging the technological affordances of the 4IR. In response to this concern, the chapter proposes the value of computational thinking as a pedagogical approach to science teaching in the context of the 4IR. It explores how computational thinking can enhance inquiry practices, problem-solving skills, and critical thinking abilities in science classrooms. The chapter also discusses the role of computational tools and technologies in supporting computational thinking within the context of science education. The chapter provides insights into the theoretical foundations and practical implications of incorporating computational thinking into science education, supporting the development of future-ready science teachers. Overall, this chapter underscores the significance of computational thinking in science education and its potential to transform teaching and learning practices in the 4IR era. It encourages educators to explore and embrace computational tools and technologies to enhance inquiry-based approaches and empower students to become proficient problem-solvers and critical thinkers in the field of science.

THE AFFORDANCES OF VIRTUAL LEARNING ENVIRONMENTS IN REMOTE CHEMISTRY LEARNING: AUTONOMOUS LEARNING, CHEMISTRY, HIGHER-ORDER THINKING SKILLS (HOTS), INDUSTRY 4.0, INTERACTIVITY

Virtual Learning Environments (VLEs) have emerged as valuable tools for facilitating teaching and learning in higher education. The research presented in this chapter aimed to evaluate the effects of VLEs on chemistry learning in remote settings. The study involved fifty pre-service science teachers from a South African university, who participated in a mixed methods approach with a quasi-experimental design. Pre- and post-tests were used to assess achievement scores for both control and experimental groups. Additionally, focus group interviews were conducted to gather reflective data. The findings reveal a positive shift in the achievement scores of the experimental group compared to the control group, indicating the potential benefits of VLEs in enhancing chemistry learning outcomes. Furthermore, the chapter discusses how VLEs can empower students to take ownership of their learning, engage in critical thinking, and interact with course materials and peers in meaningful ways.

TOWARDS AUTOMATIC ANALYSIS OF SCIENCE CLASSROOM TALK - FOCUS ON TEACHER QUESTIONS

This chapter explores the application of automatic speech recognition (ASR) for the analysis of teacher talk in science classrooms. Often educational research on classroom interactions and talk has traditionally relied on manual analysis which is a time-consuming and expensive endeavor. By leveraging ASR technology, the study aims to automate the analysis process and specifically focuses on the detection and differentiation of teacher questions, distinguishing between closed and open types.

The research data comprises ASR text outputs from 25 physics lessons, each lasting 90 minutes, taught by 25 different science teachers and centered around the same topic. By combining human-led and machine-led approaches, the study takes significant strides towards automatization in analyzing classroom talk. It examines the accuracy and efficiency of the automatic detection of different types of talk, with a specific focus on teacher questions. The findings shed light on the capabilities and limitations of ASR in capturing and categorizing teacher questions in a classroom setting. It discusses the potential of ASR-based analysis to provide valuable insights into instructional practices, student engagement, and the effectiveness of questioning strategies. It emphasizes the potential of ASR technology to streamline the analysis process and opens up possibilities for further research and development in this area.

DIFFERENT CAMERA PERSPECTIVES, DIFFERENT OBSERVATIONS? PRE-SERVICE SCIENCE TEACHERS' NOTICING AND PERCEIVED USEFULNESS ON DIFFERENT CLASSROOM VIDEOS

This exploratory study focuses on the learning experiences of pre-service science teachers (PSTs) in the context of a teacher education activity that incorporates innovative video technology. The study aims to investigate the nature of the PSTs' noticing when viewing different videos of their rehearsals, specifically focusing on the core practice of eliciting and working with student thinking. Additionally, the study explores the PSTs' perceived usefulness of viewing and reflecting on different types of video footage.

The PSTs' rehearsals were recorded using multiple cameras placed at various locations, including a video recorder positioned at the rear of the classroom, camera goggles worn by the PST, and a student actor. The chapter discusses variations in PSTs' observation foci, observation specificity, and analytic approaches when analyzing different types of video footage. The results reveal that the PSTs exhibited different patterns of attention and interpretation based on the camera perspectives used in the videos. The chapter discusses how the choice of camera perspective influenced the PSTs' ability to notice and make sense of their teaching practices. It explores the implications of these variations in noticing for teacher education and the development of pedagogical expertise.

Furthermore, the chapter examines the PSTs' perceived usefulness of viewing and reflecting on different types of video footage. Although the PSTs rated all types of footage as beneficial for enhancing their learning, they expressed more positive evaluations of videos filmed from the back of the classroom compared to point-of-view footage. The implications of using videos as a tool to support PSTs' learning and professional development are thoroughly discussed. By exploring the connections between camera perspectives, PSTs' noticing, and perceived usefulness, this chapter contributes to the understanding of how video technology can enhance teacher education. It emphasizes the need for teacher educators to consider the impact of camera perspectives on PSTs' observations and encourages further exploration of video-based approaches to enhance teacher learning and pedagogical practices.

PREPARING SCIENCE TEACHERS FOR TOMORROW WITH TRANSVERSAL COMPETENCIES

This chapter addresses the evolving needs of our education system and the importance of preparing students for the future in a society that is constantly changing. It emphasizes the necessity for agile and flexible schooling systems that equip students with transversal competencies to thrive in a world where new jobs are yet to be invented. The chapter challenges the traditional focus on content-based education and advocates for prioritizing skills development, such as critical thinking, creativity, problem-solving, communication, and collaboration. It explores how science education can play a pivotal role in developing learners who possess transversal competencies, which are essential for acquiring knowledge and engaging with science subjects. By emphasizing the importance of transversal competencies in science education, this chapter contributes to the discourse on preparing science teachers for the future. It highlights the need for education systems to prioritize skills development alongside knowledge acquisition, empowering students to curate information, think critically and creatively, solve problems, communicate effectively, and collaborate with others. The chapter provides insights into research-based approaches that foster the development of transversal competencies in science learners.

IBSE INQUIRY-BASED SCIENCE EDUCATION: AN EXAMPLE OF APPLICATION IN A PRIMARY SCHOOL

This chapter focuses on the application of Inquiry-Based Science Education (IBSE) in a primary school setting, specifically addressing the development of scientific competencies in primary school children. It describes an inquiry-oriented teaching program based on 5E method being implemented in a Grade 3 classroom.

The practical example offers insights into how the 5E methodology can be effectively implemented, providing a framework for teachers to design inquiry-based science lessons. By showcasing the application of IBSE in a primary school context, this chapter contributes to the understanding of effective science education approaches for young learners. It highlights the importance of hands-on experiences, critical thinking, and collaborative skills in fostering meaningful science learning. The chapter also emphasizes the role of inquiry-based approaches in promoting student engagement and deepening their understanding of scientific concepts. It serves as a valuable resource for educators seeking to enhance science teaching and learning experiences in primary classrooms, promoting active student involvement and the development of scientific competencies.

PRE-SERVICE TEACHERS' COGNITION OF STEM VIA 4IR COGNITIVE PRINCIPLES: COGNITION OF STEM VIA 4IR COGNITIVE PRINCIPLES

In the context of the Fourth Industrial Revolution (4IR) or Industry 4.0, the learning process and students' comprehension, whether in discipline-specific areas such as science, technology, engineering, and mathematics (STEM), or in general decision-making, can be hindered by the inappropriate application of cognitive principles. Recognizing this challenge, this chapter explores the cognitive tenets of the 4IR or Industry 4.0 and their implications for pre-service teachers' cognition of STEM disciplines.

This chapter is grounded in the theoretical framework of 4IR cognitive principles, drawing upon both theoretical and empirical research to improve pre-service teachers' understanding of mathematics and their overall decision-making skills. It explores how these principles can inform and shape the cognition of STEM disciplines, providing a theoretical framework for understanding the challenges and opportunities in this context. Additionally, the chapter presents empirical findings that shed light on the effectiveness of incorporating 4IR cognitive principles into pre-service teacher training programs. It discusses the impact of these principles on mathematical discipline cognition and general decision-making skills, highlighting the potential benefits of integrating them into educational practices. It offers insights into the theoretical foundations of 4IR cognitive principles and provides evidence-based strategies for enhancing knowledge-building and discipline-specific development.

In conclusion, this edited reference book delves into the critical theme of fostering science teaching and learning for the Fourth Industrial Revolution (4IR) and beyond. It addresses the pressing need to equip prospective teachers with the necessary competencies and skills to navigate the rapid advancements in technology and the complex challenges of the 4IR era.

Through a collection of diverse chapters, this book offers valuable insights, research-based approaches, and practical strategies to prepare teachers for the demands of the 4IR. Each chapter focuses on a specific aspect, ranging from competency-based assessments and project-based learning to virtual learning environments, classroom talk analysis, and the integration of computational thinking.

The book highlights the importance of embracing innovative pedagogies, such as inquiry-based approaches, extended reality (XR), and transversal competencies, in science teaching and learning. It emphasizes the role of technology and emerging tools in enhancing student engagement, critical thinking, collaboration, and problem-solving skills.

As editors, we believe that this book serves as a valuable resource for science educators, research scholars, and science teachers seeking to enhance their pedagogical practices and adapt to the evolving landscape of science education. It provides a comprehensive overview of key perspectives, methodologies, and tools that have promising payoffs in the lives of pre-service teachers, informing their decision-making and action.

We hope that this edited reference book inspires readers to explore innovative approaches, implement effective teaching strategies, and embrace the transformative power of science education in the 4IR and beyond. By fostering science teaching and learning that aligns with the demands of the 4IR, we can empower students to thrive in an increasingly complex and interconnected world.

We extend our gratitude to all the contributors who have shared their expertise and insights, making this book a rich and valuable compilation. Our sincere hope is that this book ignites meaningful discussions, sparks further research, and contributes to the advancement of science education for the betterment of future generations.

Together, let us embark on this journey of fostering science teaching and learning for the Fourth Industrial Revolution and beyond, shaping a future where science education plays a pivotal role in nurturing curious minds, critical thinkers, and lifelong learners.

Garima Bansal and Umesh Ramnarain

Chapter 1
Theory Into Practice:
Supporting Knowledge-in-Use Through Project-Based Learning

Tingting Li
Michigan State University, USA

Emily Adah Miller
University of Georgia, USA

Joseph S. Krajcik
Michigan State University, USA

ABSTRACT

Facing the increasingly complex and ever-changing environment of present challenges, citizens in the fourth industrial revolution across the globe will need to develop knowledge-in-use proficiencies, which equips them with the scientific knowledge to make evidence-based informed decisions, support policy changes, and understand future consequences of lack of action. However, knowledge-in-use is a complex construct with multiple different sub-components. It remains challenging to both understand the underlying information processing mechanism and the inner cognitive process of knowledge-in-use. This chapter unfolds the latent process of developing knowledge-in-use by referencing several landmark cognitive models. Building on that, this chapter discusses how a project-based learning environment has the potential to support students' knowledge-in-use development.

DOI: 10.4018/978-1-6684-6932-3.ch001

INTRODUCTION

Facing the increasingly complex and ever-changing environment of present challenges, such as food scarcity, pandemics, and climate change, citizens in the fourth industrial revolution (4IR) across the globe will need to equip themselves with scientific knowledge to make evidence-based informed decisions, support policy changes, and understand future consequences of lack of action (Anderson, 2020; National Research Council, *hereafter* NRC, 2012a; NRC, 2011; Organization for Economic Co-operation and Development, 2016). To develop such a science-literate citizenry, educators must consider what students should ultimately know (big ideas) and be able to do (scientific practices) and how to develop learning environments to support this integrated type of proficiency (NRC, 2012a). As such, the goals of science education around the world have shifted to knowledge-in-use learning goals (Finnish National Board of Education (FNBE), 2015; Germany (Kulgemeyer & Schecker, 2014); USA, NRC, 2012a, 2012b; People's Republic of China Ministry of Education, 2003). Knowledge-in-use requires all students to demonstrate and apply their knowledge, rather than recall what they know, by making sense of real-world phenomena, solving complicated problems and making informed decisions (NRC, 2012a; National Academies of Sciences, Engineering, and Medicine [NASEM], 2019; Pellegrino & Hilton, 2012).

Knowledge-in-use allows students to explain novel real-world phenomena or solve complex problems by applying what they learned (NRC, 2012b; NGSS Lead States, 2013). While the critical role of knowledge-in-use has been recognized and continues to receive critical attention (Charara et al., 2021; Harris et al., 2019; He et al., 2023; Pellegrino, 2012), it is a complex construct with multiple sub-components. As a result, it remains challenging to understand both the underlying information-processing mechanism and the underlying cognitive process of knowledge-in-use. Moreover, without understanding the nature of knowledge-in-use, it is even more challenging to explore further how to support an individual's development of knowledge-in-use effectively.

From a cognitive and psychological perspective, knowledge-in-use maps onto the core characteristics of adaptation (including transfer, flexibility, cognitive flexibility, creativity, and cognitive rigidity) (e.g., Lussier, Shadrick, & Prevou, 2003; Shadrick, Crabb, Lussier, & Burke, 2007; Shadrick & Lussier, 2004; Spiro et al., 2019; Ward et al., 2018). All characteristics emphasize the applicative, social, and interactive nature of learning in multiple contexts with different affordances (Broudy, 1977; Greeno, Smith, & Moore, 1993; Lave, 1988). These characteristics resonate with the main features of project-based learning (PBL) (Blumenfeld et al., 1991; Krajcik & Shin, 2014). Grounded in learning theories, PBL is a comprehensive approach to classroom teaching and learning designed to engage students in investigating authentic problems, which guides designing learning environments. The PBL approach triggers student interest and enables students to develop and apply usable scientific knowledge to make sense of compelling and complex phenomena or address complex problems (Inkinen et al., 2020; Schneider et al., 2016). Substantial evidence indicates that students who engage in PBL classrooms outperform students in non-PBL classrooms on both science academic achievement and engagement from elementary school to high school (Geier et al., 2008; Harris et al., 2015; Krajcik et al., 2023; Schneider et al., 2020; Schneider et al., 2022; Shin et al., 2019). Nonetheless, to date, limited research focuses on how to design PBL environments to develop students' knowledge-in-use from the view of underlying cognitive processes, which is an urgent need in the 4IR.

Two primary goals guide the writing of this chapter. First, we unfold the latent process of developing knowledge-in-use by referencing several landmark cognitive models. Our second goal for this chapter is to discuss how a project-based learning environment has the potential to support students' knowledge-in-use development. We aim to provide insight into the underlying cognitive process of knowledge-in-use. Based on that aim, we further explore how to support individuals in developing knowledge-in-use for solving novel problems.

The outline of the chapter is as follows: We start the chapter by discussing the notion and characteristics of knowledge-in-use while expounding on the critical position of knowledge-in-use for the new science education reform vision, which is needed for the 4IR. Then, to inform knowledge-in-use's understanding and development, we review some seminal cognitive models in cognitive science and principles for supporting adaptive skill development. Next, we discuss how PBL provides affordances for supporting student knowledge-in-use development from theoretical perspectives. We also present examples of our developed PBL environment to offer insights into designing learning environments to promote useable knowledge. Finally, we summarize the features of the PBL environment and propose a future research agenda.

New Vision of Science Education in the Era of 4IR: Knowledge-In-Use

Next, we discuss the core notions of the latest science education reforms in the US, which are also the critical ideas of the global science education reform.

Building on learning theories of situated cognition (Greeno et al., 1996) and social constructivism (Vygotsky, 1978), the USA policy document, *A Framework for K-12 Science Education* (NRC, 2012) put forth a new vision of science proficiency—defined by others as knowledge-in-use, which all students should develop (Pellegrino & Hilton, 2012; NRC, 2012b). The knowledge-in-use perspective reflects the increased awareness among learning scientists, science educators, and policymakers of thc proficiencies required for global citizens in the 21st century (OECD, 2019). Knowledge-in-use argues that knowledge is a product of the activities, context, and culture in which it is developed and used (Brown et al., 1989; Bonwell & Elson, 1991) and posits that individuals are active participants in the creation of their own knowledge (Schreiber & Valle, 2013). Therefore, rather than acquiring knowledge from teachers or textbooks, knowledge-in-use emphasizes the use and application of scientific knowledge to make sense of natural phenomena and provide solutions to complex, authentic problems, promoted in *The Framework* (NRC, 2012b).

Notably, a phenomenon (e.g., the process of evaporation) or a complex problem (e.g., maintaining freshwater resources) is always multi-dimensional and requires students to selectively use a range of ideas (e.g., properties and structure of matter, interaction, and energy, earth systems) from different domains. Moreover, students should apply those selected ideas to new contexts. However, knowledge-in-use cannot be developed instantaneously; instead, it develops over time and requires deep learning when students make connections among ideas and integrate various experiences (Adah Miller, et al. 2022; Esposito & Bauer, 2017; Miller & Krajcik, 2019). In other words, this vision for knowledge-in-use requires sustained immersion in a learning environment that promotes deep learning by integrating core science ideas and practices (NRC, 2012). To this end, knowledge-in-use is consistent with other high-order thinking skills, such as adaptive skills, transferable knowledge, and cognitive flexibility (Spiro et al., 2019; Ward et al., 2018). We cannot expect to encounter cases that are identical in real life; only through transferring knowledge from similar contexts can we solve complex and unique, real-world problems.

The efficient way to cope with the real-world problem should be tailoring various aspects from individuals' prior experiences, including knowledge or skills, then integrating and accommodating those tailored aspects to the new context for problem-solving. Learners with knowledge-in-use should be able to selectively incorporate ideas from a range of knowledge domains and apply them flexibly in varied and demanding ways (NRC, 2012a; NGSS Lead States, 2013). However, question remain: How can we support individuals in developing knowledge-in-use in the novel context? This has been a central question and one of the most challenging, contentious, and critical questions in psychology and education (Day & Goldstone, 2012), Its resolution is critical to prepare individuals for 4IR. To answer this question, we must unfold the cognitive process of knowledge-in-use.

Theoretical Approach to Knowledge-In-Use: Adaptivity Models

We share a common understanding of knowledge-in-use with adaptivity research. Thus, in this section, we mainly discuss several cognitive models of adaptivity. By adaptivity, following Albert and Haves (2003), we mean "the ability to employ multiple ways to succeed and the capacity to move seamlessly among them" (Hoffman, Ward, et al., 2014, pp. 51-52). Subsequently, we briefly summarize the major adaptivity research and move to several landmark cognitive models of adaptivity.

Adaptivity has a long tradition of research (e.g., Ward et al., 2018). Throughout its history, a somewhat confusing array of terms have been used to refer to skilled adaptation of disparate kinds. Among them, adaptive expertise (Hatano & Inagaki, 1984; 1986) has received wide consensus on the origins of experts' quality of being adaptive (Ward et al., 2018). Individuals who have adaptive expertise cannot only perform procedural skills efficiently but also have a strong conceptual understanding of those procedures and the associated contexts expertise (Hatano & Inagaki, 1984; 1986). Adaptive expertise contrasts with routine expertise, which is often described as "outstanding in terms of speed, accuracy, and automaticity of performance" (Hatano & Inagaki, 1984; p. 31). Routine expertise is limited to near transfer, which often lacks flexibility and adaptability to new problems (Hatano & Inagaki, 1984; p. 31). Conversely, according to Hoffman (1998), "experts are adaptive by definition." Building on that, Hatano and Imagaki (1986) asserted that adaptive expertise requires a conceptual understanding of procedural skills that afford in-event adaptive thinking.

When exploring the mechanism of adaptive expertise, however, most explanations of adaptive expertise are similar to the broader mechanisms proposed by others to explain expertise (Chi, Feltovich, & Glaser, 1981). Various cognitive models of adaptive expertise have been proposed, all emphasizing the critical role of sensemaking and flexibility in developing adaptivity. In the next section, we focus on several related seminal works.

Data-Frame (D/F) Model of Sensemaking

Klein and his colleagues (2006) proposed and designed the data-frame (D/F) sensemaking model (Figure1) to further unpack the diagnosis loop featured in the earlier related model of "recognition primed decision" (RPD; Klein, Calderwood, & Clinton-Cirocco, 1886; Klein, Calderwood, & MacGrego, 1989). The diagnosis loop in RDP focuses on the key cues for decision-making, reasonable goals and expectations that enable individuals to understand the evolving situations and ultimately serve the pursuit of a course of action. The D/F sensemaking model treats sensemaking as a process of adaptive understandings (Ward et al., 2018).

Figure 1. D/F model of sensemaking
(adopted from Klein et al., 2006)

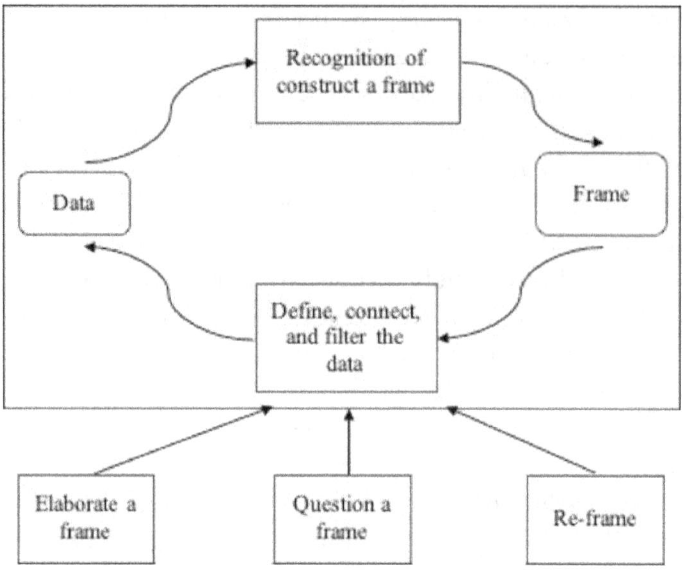

The D/F sensemaking model presents the pathways for elaborating, questioning, or reframing the situations that we encounter. Klein et al. (2006) proposed the model of sensemaking, focusing on how individuals make situational assessments in the face of an anomaly or surprise. Rather than pursuing a course of action, the D/F model illustrates situational understanding and situational assessment mechanisms. Based on the data available in that situation, individuals should construct a frame that represents an explanatory structure showing the relationships among various entities. The frame is an interpretation of the data, and it serves as a lens to guide the additional data searching to refine the frame iteratively. This model provides a context-sensitive mechanism of how we understand a current situation and illustrates how to cope with a new situation by either questioning or elaborating on the frame constructed or by coming up with an entirely new frame (reframing process). The D/F model explains how people understand a given situation by combining available information and significant clues. However, as stated by Ward et al., viewing the relationship between data and frame as reciprocal presents difficulties in the sense-making process. Indeed, the real challenge for sensemaking is to build a frame based on various data and situational cues. However, there are still potential risks in creating a template-based model, even if these frames can be applied to some well-structured domain. Additionally, the frame used in one situation cannot be transferred to another, especially in complex and poorly defined domains. For instance, individuals cannot expect to use exactly the same problem-solving schema that worked in one context to solve a different problem in a different context with different contextual features.

Macrocognitive Model (D/F +F) of Adaptive Skill

To address the issue of an ill-structured domain, Hoffman and colleagues (2014) integrated two models that explain the adaptive process from two perspectives (sensemaking and replanning) into an integrated D/F+F model (see Figure 2, below). They sought to better illustrate the process of conceiving adaptivity. The model includes two components, data-frame (D/F) sensemaking model (Klein, et al., 2006) and Flexecution (F) model of adaptive planning. During employing skilled adaptation, sensemaking, and replanning interact in the model and are interdependent with each other. Ward et al. (2018) also underpinned both sensemaking and flexible execution as critical to leadership and the ability to adapt. Sensemaking means an individual should first make sense of the phenomena using data to generate a frame for understanding the phenomenon. Flexible execution (flexecution) means how individuals flexibly pursue goals in the types of complex settings that are characterized by emergent and unpredictable challenges (Klein, 2007a, 2007b, 2011).

Compared with the D/F model, which mainly focuses on the understanding or sensemaking processing of new situations, the F model of adaptive planning, on the right side of Figure 2, describes how individuals flexibly pursue goals in complex environments with emergent and unpredictable challenges (Klein, 2007a, 2007b, 2011). Flexeution, a model of adaptive planning, illustrates "a management-by discovery approach where both courses of action and goals are adapted to meet the operational demands of goals that emerge dynamically, some of which conflict, are reprioritized, juggled, traded-off, or become obsolete (Klein, 2011)." This model recognizes that goals change as actions unfold and that plans must be flexible in their ways, means, and ends. In this model, action is a vital term as it emphasizes how the original plan is adapted to an evolving situation. Action provides both challenges (emergent anomalies) and opportunities (new information or idea) to help make an adaptive decision. Elaborating the action plan, questioning the goals, and reframing the priorities are three ways individuals can evaluate their initial objectives and action plans. This process gives individuals sufficient learner control. Like the D/F model, the critical step in this model is how individuals evaluate both the initial plan and the goals to adapt the plan. Nevertheless, although the F model provides three ways to achieve it, there is still some room for amplification in adapting the plan, especially in the complex domain. That is, for most real-world phenomena or problems, which are often ill-structured, they require iterative and adjusted processes to figure things out. To this end, more fine-grained sensemaking and felxecuting processes are needed to guide knowledge or experience transfer processes.

Figure 2. An integrated (D/F+F) macrocognitive model of adaptive skill (adopted from Hoffman, Ward, et al., 2014)

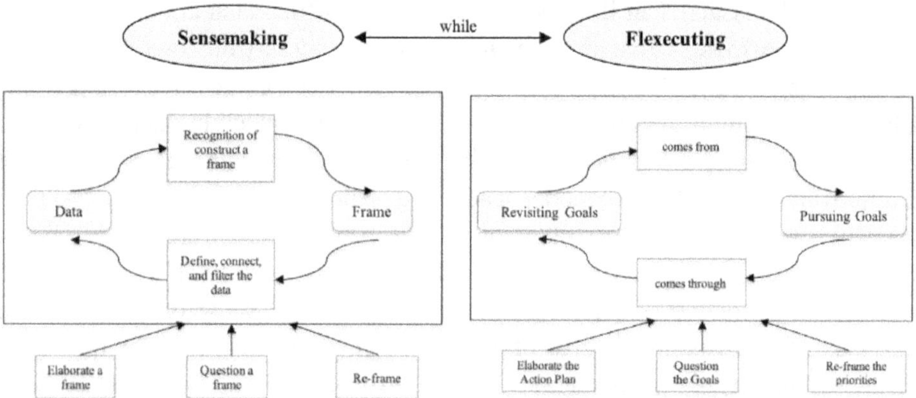

Microcognitive Model (D/F +F) of Adaptive Skills

The knowledge-in-use perspective emphasizes real-world situations, which are often complex and ill-structured. As discussed before, the D/F model of sensemaking offers us a possible way for individuals to understand situations. However, the nature and process of the frame construction need to be further unfolded and elaborated, especially for new situations that require the integration of current contextual information with previous experience. Ward et al. (2018) points out, "studies are needed to confirm the hypothesis that adaptive framing is central to expertise in complex real-world contexts." To delve deeper into this matter, we focus on the sense-making and adaptive planning processes and examine the factors that may impact the framing. We modified the macrocognitive model (D/F + F) of adaptive skills into a microcognitive model (D/F+F) to unfold the sensemaking process.

We argue that individuals make sense of emergent situations and adapt their initial plan based on sensemaking in an ill-structured domain, complicated real-world situation, and "ordinary novelty." In the well-structured domain, schema theories are limited by their reliance on preformed knowledge structures retrieved from long-term memory and instantiated by features of some present context (Spiro et al., 2020). Considering the dynamic, non-linear, and complex features of an ill-structured domain, we draw on the Cognitive Flexibility Theory (CFT) to address the importance of the deliberate practice of open and context-dependent meaning-making. Meaning-making enables concepts to be ready to be adapted to new contexts and have greater readiness to participate in situational sensemaking (Spiro et al., 2020). CFT focuses on the adaptive processing of non-routines to entail a shift from retrieving intact structures to assembling aspects of the prior case and knowledge structure to fit the context-sensitive needs of the situation at hand. The novel contexts, which occur with great frequency in the unconstrained real world, are the primary context that CFT considers.

According to CFT, reusing prior ingredients in novel assemblies and deploying common structural meta-features of novel complex cases to organize in an adaptive worldview are two solutions that can support individuals to adapt their prior experience to new situations (Spiro et al., 2020). Individuals can gain conceptual variability through deliberate practice in various cases. Conceptual variability has the features of conceptual understanding (context tailored) and context dependent. When there are more cases or situations that an individual experiences in the learning process, the more complicated it becomes, and conceptual variability interacts forming concept-combination variability. CFT refers to the general conceptual-variability feature as an adaptive worldview. When encountering a novel situation, the readiness of concepts combines with other concepts as adaptive assemblages, that is, schemas of the moment, which fit the novel context. In this process, individuals construct two different types of schemas, one is specific content schemas (ontological features) without general applicability, and the other is a common set of meta-features of an adaptive mindset (epistemological features) found across all situations, which is a general approach that can be used adaptively.

Considering the unique characteristics of ordinary novelty situations and CFT, we argue that an individual's prior experience of concepts in different contexts exerts a vital influence on their sensemaking of emergent situations, which further affects his or her adaptive skills. We modified Hoffman et al.'s integrated macrocognition model of adaptive skills to address and elaborate the sensemaking and replanning processes with the participation of various prior experience structures (see Figure 3).

Figure 3. Microcognition model of adaptive skills

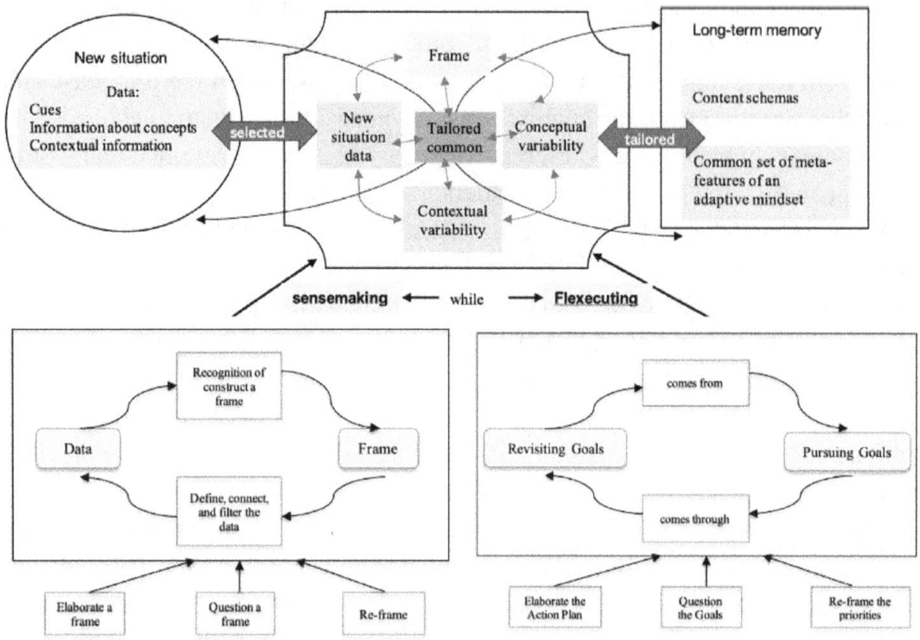

In the microcognitive model of adaptive skills, we emphasize the process of constructing the adaptive worldview across various contexts. We also highlight the interaction between the adaptive worldviews and new context to generate schema for the moment.

As shown in the model, when individuals encounter novel situations, they will first evaluate the new context. Then, based on the data and cues that the context provides, they recall various cases or situations they have experienced. Through the commonalities and differences analysis among the prior contexts and the new context, individuals will select the prominent parts of each context. They pick up a common meta-feature of an adaptive mindset and specific content schemas from their long-term memory. The common epistemological aspect of the adaptive worldview and the specific content features will be further tailored to fit the context information. The tailored epistemological adaptive worldview interacts with the tailored specific content schemas to construct a frame for the novel context. In so doing, one can make sense of the novel situation and can evaluate the initial plan for adapting to their new decisions.

The microcognitive model of adaptive skill elaborates on the specific process of how adaptive thinking happens when one encounters a new context. Prior experience in various cases or contexts plays an essential role in this process. However, further questions remain: how do individuals select contexts? How can we deliberately develop an individual's adaptive skills? Ward et al. (2018) proposed six principles to apply in order to develop adaptive skills.

Cognitive Perspectives of Adaptive Skills Training Principles

After a systematic review of existing literature and related theories, Ward et al. (2018) presented an argument that corresponds to the point about adaptive skills of Flexecution model. Ward and others argue that, "Fundamental to the achievement of adaptive performance is the opportunity for practice at problems that stretch current competency," The flexecution model shows that adaptive performance is achieved by doing; it can be termed performance by discovery. Deliberate practice of doing in differing situations or contexts can be one possible way to develop individuals' adaptive skills. These practices are needed to revise and extend one's conceptual understanding, acquire new knowledge, reasoning skills, and reformulate knowledge and skills (adaptive worldviews) to fit different situations or contexts. Ward et al. also proposed six adaptive skill educational principles based on the existing work. These training principles are:

- Principle 1: Flexibility-Focused Feedback. This principle asks to provide feedback that promotes cognitive flexibility and allows learners to overcome the human tendency to stick with current methods (e.g., generate reasons not to change) (e.g., Klein & Baxter, 2006; Feltovich et al., 1997). For instance, if an individual has tried to solve a problem, flexibility-focused feedback should allow the individual to reflect on what strategies they have used so far for similar problems, when the current strategies work effectively, and when they will not. The feedback should also promote new strategies development that permit them to demonstrate effective flexibility and adaptations to new situations with unexpected variants from prior experiences.
- Principle 2: Concept-Case Coupling. Learners should be provided with opportunities to experience the full variation of ways a given concept is manifested across particular cases. This principle allows to reconnect cognition and context, especially facilitating an experience of how concepts vary in situ from case to case and allows learners to exposure to both situational and action variants of any given concept and their reciprocal effects. For instance, the phenomena or cases that individual experiences should be able to have either conceptual or contextual unique affordances for the individual to develop adaptive skills.
- Principle 3: Tough Case Time Compression. This principle addresses time compression via a test bank of severe cases and provides the opportunity

to increase the frequency with which learners can practice, especially at coupling concepts with cases and variants of those cases. That is, the cases we provide for learners should try to keep their authenticity (e.g., dilemmas in the cases, challenging or complex cases) and increase a learner's exposure to the gamut of tough cases in all their guises.

- Principle 4: Case-Proficiency Scaling. Training should ensure cases are tailored to current needs and proficiency levels to stretch performance just beyond that which is currently attainable. Specifically, for adaptive skills, training should leverage training content form expert mental models, scale tough cases to each specific proficiency level using the zone of proximal development and maintain training at the edge of the zone of proximal development.
- Principle 5: Complexity Preservation. This principle is to avoid the trend that oversimplifies the complexity of the situations in which learning takes place. It calls for preserving the situation's complexity. To do it, learners should practice in varied contexts and receive opportunities to adapt to complexity, novelty, and changing demands and goals.
- Principle 6: Active Reflection. Metacognition and reflection play an essential role in developing adaptive skills, which should be emphasized before, during, and after learning. Training or learning should help learner calibrate their own understanding of their skills, knowledge, and understanding by predicting or estimating their competency, performance, or learning. Learning should also allow an individual to elaborate on their current understanding of their own representation of the problem space by engaging in experimentation with new cases; extrapolation of experience with prior incidents to new incidents; and explanation of what happened, why, and how success was achieved.

Although the field of learning research have gained more knowledge of how to educate individuals to develop adaptive skills, there remain key challenges. For example, we have yet to understand how we can incorporate these principles into instruction to develop learners' adaptive skills. Several studies address this question and indicate that the possible solution is learning in a real-world context without oversimplifying (e.g., Adah Miller, et al., 2022; Spiro et al., 2020; Day & Goldstone, 2011). These framings critique cognitive learning models (employed in schools) as decontextualized, as well as individualized pedagogical models which describe knowledge as an independent possession of the learner (Wilson & Myers, 2000). That concepts are abstract, self-sufficient, and separate from the context does not align with the idea that learning is an active and engaging process (Bonwell & Eison, 1991; Brown et al., 1996).

Furthermore, acquiring abstract knowledge does not guarantee transferable knowledge – this is similar to a situation when a person has quality tools but cannot use them. The gap created between knowing and doing prevents learners from using their knowledge within an appropriate context (Bransford et al., 2000; Brown, Collins, & Duguid, 1989). In terms of the instructional problem, situated learning theory (Lave, 1988) receives wide attention (Cobb & Bowers, 1999). It integrates most of the separate branches of an investigation into a complete theory of learning and transfer. In this view, learning and cognition are situated and developed through purposeful, authentic activities in social contexts. Therefore, learning and transfer occur when learners are allowed to "observe and practice in situ" (Brown et al., 1989, p.34).

PRACTICAL APPROACH TO DEVELOP KNOWLEDGE-IN-USE: PROJECT-BASED LEARNING

One of the most efficient ways to assist learners in adapting to the demands of novel situations is to provide them with the appropriate knowledge and skills necessary to confront and solve complex real-life problems (Brown & Duguid, 1993). One effective practical approach to implement this idea is to design learning environments that offer authentic activities for novices to experience expert performance, provide scaffolding at important moments, offer adequate support for cooperative knowledge building, and have monitoring features within the learning environment. (Herrington & Oliver, 1995, 2000). As one of the inquiry-based instructional approaches, project-based learning (PBL) creates an authentic platform for integrating three-dimensional learning to promote students' knowledge-in-use (Miller & Krajcik, 2019).

Project-based learning (PBL) (Blumenfeld et al., 1991; Krajcik & Shin, 2014) is a student-centered learning model that immerses students in meaningful questions to collaboratively design physical artifacts. This approach responds to the research question by engaging students in long-term meaningful projects. PBL provides a more inclusive environment than a traditional curriculum by purposefully leveraging student interest. PBL focuses on an asset approach to teaching and learning. It can provide learning opportunities for students across all backgrounds and demographics (Krajcik & Czerniak., 2018; Miller & Reigh, 2020). Next, we first present an overview of PBL. Then, we discuss the theoretical underpinnings of PBL. Finally, we discuss the main features of PBL and project-based learning environments.

Overview of PBL

PBL is an approach to classroom teaching and learning designed to engage students in investigating authentic problems. Even though there are different versions of PBL (Barron, et al., 1998; Krajcik et al., 1998), they all hold some common features. First, PBL uses a driving question that is meaningful to learners to drive their exploration and sustains motivation across time. Second, the exploration and investigation are driven by questions that can result in concrete artifacts that respond to the driving question through a culminating learning sequence. Last, the driving question and the resulting artifact the students create from the investigation are connected authentically with the students' community (Miller & Krajcik, 2019; Helle, Tynjälä, & Olkinuora, 2006). PBL units are relatively long-term, problem-focused, and meaningful instruction experiences that integrate principles and concepts from several disciplines or fields of study.

Given its focus on students and their interests, PBL tailors to the intellectual resources and experiences of diverse students and is responsive to culture, race, SES, and gender (Boaler, 2002; Geier et al., 2008; Krajcik et al., 1994). Teachers and communities also agree that PBL can motivate students to learn (Beneke & Ostrosky, 2009; Chu, Tse, & Chow, 2011). School-wide PBL models can be found across the United States and many other countries such as Finland, Germany, Israel, and Denmark (Schneider et al., 2016; Tal & Miedijensky, 2005; Thomas, 2000). However, PBL models have not resolved scale and sustainability questions, especially for less well-funded public schools (Coburn, 2003; Potvin, et al., 2022). Furthermore, there are challenges of scale, sustainability, and persistence of PBL, especially integrating PBL with reform-based standards or top-down initiatives (Coburn, 2003; Halvorsen et al., 2012; Potvin, et al. 2022).

PBL has been applied in various disciplines, such as social studies, science, technology, and increasingly for literacy and mathematics. It is also tailored to meet the specific practices associated with these disciplines (Bell, 2010; Boaler, 2002; Krajcik & Mun, 2014). The incorporation of PBL approaches across disciplines has begun to be widely developed. In terms of PBL in science education, as one of an inquiry-based instruction approaches, PBL emphasizes engaging students in using scientific and engineering practices to explain phenomena and find solutions to real-world problems. These practices are similar to those of professional scientists, which are collaborative and can increase student engagement. Practices that mirror the practices of scientists help students develop a deeper understanding of disciplinary core ideas and crosscutting concepts (Miller & Krajcik, 2019). The Next Generation Science Standards calls for promoting science learning as those understandings and capacities for using the knowledge that develops over time in a coherent manner. Three-dimensional (3D) learning is considered one approach to support student knowledge-in-use development (Miller & Krajcik, 2019). 3D learning means learning occurs by integrating disciplinary core ideas, scientific and engineering practices, and crosscutting concepts to explain real-world phenomena. 3D learning builds on prior experiences with ideas, practices, and crosscutting concepts over and across the years. Through a dynamic process of investigation and collaboration, students are motivated to sustain the inquiry and explore a relevant topic. Evidence indicates that students who engage in a project-based science curriculum develop a deeper understanding of science content and processes than traditional approaches, such as textbooks and lectures (Fortus et al., 2004; Geier et al., 2008; Harris et al., 2015; Marx et al., 2004; Schneider et al., 2020; Shin et al., 2019; Tal, Krajcik, & Blumenfeld, 2006; Krajcik, et al, 2023).

Despite the prevalence of PBL curricula, PBL remains under-researched (Halvorsen et al., 2012), especially in elementary education. A major call for the field has been to gather evidence that PBL environments correlate with student learning of knowledge-in-use standards in rigorous gold-standard randomized control trials (Kokotsaki et al., 2016).

The Theoretical Foundation of PBL

Investigating real-world, meaningful questions has long been recognized as a viable instructional method (Blumenfeld et al., 1991; Krajcik & Czerniak, 2018). PBL combines Dewey's ideas with insights from learning science research (Bransford, Brown, & Cocking, 1999; NRC, 2012). It can lead to students' increased engagement in science learning (Schneider, Krajcik, Lavonen & Salmela-aro, 2020). Active construction, situated learning, social interactions, and cognitive tools are four ideas that PBL is grounded in (Krajcik & Shin, 2014). Besides these four ideas, to meet the challenges of diversity, universal learning design is also added as the fifth idea in this manuscript. Various types of scaffolds and universal learning design principles – multiple forms of representation, multiple means of expression, and multiple means of engagement are provided (Rose, Meyer & Hitchcock, 2005).

Situated Learning. Situated learning is one of the three general approaches of knowing and learning (Greeno, Collions, & Resnick, 1996). This theory argues that knowledge is situated, being in part a product of activities, context, and culture in which it is developed and used (Brown, Collins, & Duguid, 1989). The situated perspective on knowing focuses on how knowledge is distributed in the world among individuals, tools, artifacts that participants use, and communities and practices in which they participate. From a situated view, learning becomes attuned to constraints and affordances through participation, and the performance of transfer in a new situation depends on how the learner responds to the constraints of activity in the current situation. For the transfer to be possible, there must be some constraints and/or affordances invariant under the transformations that change the learning situation into the transfer situation. That is, if a transfer occurs, then there must be something (e.g., prior experience that informs the new case) constant between the original and the new case. For a transfer to occur, the learner must become attuned to those invariants in their initial learning (Greeno, Collions, & Resnick, 1996). From a situated perspective, learning involves improving one's ability to participate in cognitive systems that are distributed and focused on involvement, which helps to maintain personal relationships. It involves authoring identity in communities in which the person participates or involves satisfying interactions with environments in which the individual has a significant personal investment. In designing a learning environment, situated learning proposes that the learning environment can be organized to foster learners' participation in inquiry. Further, learning practices support the development of learners' personal identities as capable and confident learners and knowers. In developing curricula, situated learning theory argues that sequences of learning activities can be organized with attention to students' progress in a variety of disciplinary practices of discourse and representation. Learning activities can focus on problematic situations that are meaningful in terms of students' experience and on how concepts and methods of subject-matter disciplines are embedded.

Learning sciences research (NRC, 2012) has shown that the most effective learning occurs when learning is situated in an authentic, real-world context. In some scientific disciplines, scientists conduct experiments in laboratories; in others, they systematically observe the natural world and use observations to draw conclusions. Situated learning in science involves students experiencing phenomena as they take part in scientific practices such as designing experiments, constructing explanations, developing models, and expressing ideas to others. A benefit of situated learning is that students more easily see the value and meaning of the tasks and activities they perform (Krajcik & Shin, 2014). When students follow detailed steps in the textbook to do a science activity, they are not significantly more active than passively listening to a lecture. In both cases, students struggle to see the meaning of what they're doing. However, when they create their own experiments designed to answer a question that they helped to frame, and which is important to them and their community, they see how science can be applied to solve important problems. In such environments, students develop an integrated understanding in which ideas are richly connected. Situated learning provides a rationale about why engaging students in authentic activities supports learning, and how learning happens through making sense of a phenomenon. Learners find meaning in those activities. It builds a foundation for the learning environment.

Active construction. Active construction enables students to construct meaning based on their experiences and interactions in the world to improve their integrated understanding (Bransford, Brown & Cocking, 1999). When students make sense of phenomena, they try to figure out why a phenomenon occurs and design solutions to problems. They make connections among ideas that they have and new ideas about what they are experiencing. In this way, students enrich their knowledge structures and acquire integrated understanding. However, the process of building these cognitive structures often takes time, and it requires a significant effort from students, particularly when it comes to completing tasks. Learners need to find the tasks that they work on meaningful, and the tasks must necessitate synthesizing (Krajcik, Palincsar, & Miller, 2015). Focusing on critical ideas in depth enables students to connect those fundamental ideas and principles, deepen their integrated understanding, and then apply their understanding to novel situations (Fortus & Krajcik, 2011).

Social interaction. Social interaction plays an essential role in students' learning (Sawyer, 2014). Questioning, clarifying, analyzing, and debating ideas with others can afford opportunities for students to develop their ideas and principles of understanding (Blumenfeld et al., 1996). Doing science requires students to be co-constructors of understanding the natural world through collaborative, discourse-intensive activities. That's also what science and engineering practices call for. Students engaged in scientific practices collaboratively become a critical part of doing science and improve both their science understanding and content-specific knowledge.

Cognitive tools. Cognitive tools also play an essential role in supporting students' science learning. Providing students with appropriate scaffolds can allow them to carry out tasks that are not possible without this assistance and support (Edelson & Reiser, 2006). Technology tools provide various scaffolds, such as allowing learners to access and collect a range of scientific data and information that cannot be easily collected via hands-on experiments. The technology tools also provide online portals/ platforms for students to plan, build, and test models to explain real-world phenomena or solve complex problems. Additionally, technology tools provide innovative assessments to better gather information to track students' learning progress and provide timely and customized feedback for each student. Some multimedia or online platform can also provide a safe online community for students to interact, share and critique the ideas of others (Krajcik, Palincsar & Miller, 2015).

PBL Design Principles/ Features

PBL is a form of situated learning, drawing on the constructivist theory that a deep understanding of the material is gained when students actively construct their understanding by working with and using ideas in real-world contexts. Students encourage learning by doing, apply ideas, and solving problems. They engage in and persist in finding solutions to real, meaningful problems that are important and interesting to them. In so doing, students attain better learning outcomes than students attain in traditional classrooms (Geier et al., 2008; Marx ct al., 2004; Rivet & Krajcik, 2004; Williams & Linn, 2003). There are six key features or design principles for a project-based learning environment: driving questions, learning goals, engaging in scientific practices, collaboration, learning technologies, and artifacts (Blumenfeld et al., 1991; Krajcik et al., 1994; Krajcik & Czerniak, 2018; Miller & Krajcik, 2020) (see Figure 4 below).

Figure 4. PBL key features and relationships

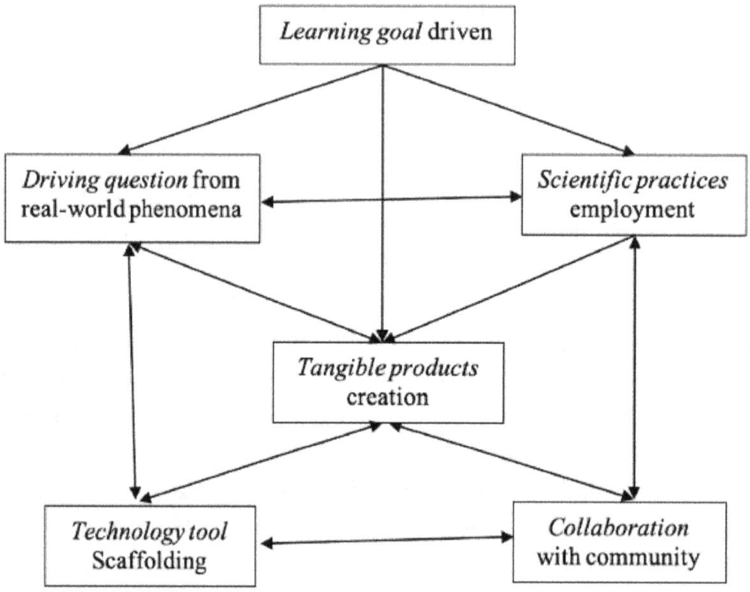

- Feature 1: Start with a driving question about a real-world phenomenon or engineering problem to be solved that promotes wonderment about the world. All driving questions in a unit can be organized on a driving question board to represent students' cognitive trajectories.
- Feature 2: Focus on integrated performance-based learning goals to demonstrate mastery of essential science standards that integrate core ideas and practices.
- Feature 3: Explore the driving question by participating in scientific practices and processes of problem-solving central to expert performance in the discipline.

- Feature 4: Explore the driving question by engaging in collaborative sensemaking activities to engage in shared knowledge building – the solutions to the driving question. This mirrors the complicated social situation of expert problem-solving.
- Feature 5: While engaged in science practices, learning is scaffolded with technologies or discourse tools that help them participate in activities usually beyond their ability.
- Feature 6: Over time, students iteratively and cohesively create a set of tangible products that scientifically address the driving question with increasing sophistication. These are shared artifacts, publicly accessible external representations of the class's learning with local impact.

Developing Knowledge-In-Use Through Project-Based Learning Environment

We discussed above the microcognitive model of adaptive skills, adaptivity training principles, and PBL features. In the following sections, we discuss why a project-based learning environment has the potential to support students' knowledge-in-use development.

Complexity Preservation Through Engaging in Real-World Contexts

In PBL, students engage in exploring real-world phenomena and engineering problems that are meaningful to them. The driving question, the hallmark of PBL, guides instruction (Krajcik & Czerniak, 2018; Krajcik & Shin, 2014; Oyewo, et al., 2022). Meaningful questions or problems is essential to PBL (Blumenfeld et al., 1991). Anchoring in a real-world situation that is meaningful and important to students, the driving question serves to organize and drive activities of the project, provides a context in which students can use and explore the phenomenon or problem, and continuity and coherence to the full range of project activities (Krajcik & Shin, 2014). Contextualization is one of the most important criteria of the driving question, which means the that students see the driving question as real-world and important. In addition, the driving question must be feasible so that students can design and perform investigations to answer the questions; it must be meaningful, interesting and exciting to motivate learners and engage them in the activities; and it must be worthwhile -another critical feature. When students generate driving questions by themselves, it is challenging for them to develop questions that contain rich science content, that meets essential learning goals, and relates to what scientists do (Blumenfeld et al., 1991; Krajcik & Shin, 2014).

Blumenfeld et al. (1991) argued, "The prevalence of low-level tasks contributes to students' lack of understanding of content and process and poor attitudes toward learning and schooling. The introduction of more cognitively complex tasks, which provide opportunities for solving real problems, often is urged as a remedy for this situation." The phenomena students engage in PBL are real-world, complex situations that preserve complexity. Engaging in these real-world, highly complex, and dynamic domains can effectively push learners to understand how to make productive sense of phenomena that do not accurately reflect how they manifest in the real world.

Take one of our project's third grade PBL curriculum units as an example. The unit driving question is "*Why do I see so many squirrels, but I can't find any stegosauruses?*[1]" This is an authentic and complex phenomenon. Units are subdivided into Learning Sets framed by more specific questions that build toward a response to the unit DQ. Each Learning Set represents a coherent sequence of lessons that lasts 1 to 2 weeks. In each lesson, students use one or more of the eight SEPs as well as DCIs and CCCs as outlined in the *Framework of K-12 Science Education (NRC, 2012).* As students' progress to each new lesson and Learning Set, their understanding of the unit phenomenon gradually deepens. For instance, in the Squirrel Unit, to make sense of the phenomenon, students engage in the long-term unit that includes 6 Learning Sets with 29 lessons. Their learning time represents the equivalent of 39 fifty-minute class sessions. After completing the unit learning, students will gain an understanding of the following enduring understanding: "*Students will apply the following ideas to explain phenomena or design solutions to problems they experience in their environment: Species can survive hundreds of millions of years because organisms in that species are either very adaptable to changes in the environment (including changes in climate and changes in the other organisms) or because of wide diversity among them. We know about some extinct species by recreating prehistoric environments using fossils and what we know about extant organisms.*"

Concept-Case Coupling Through Various Contexts of Organization

Hierarchical relationships among various phenomena within a PBL unit. The phenomena and problems in PBL are the drivers of increasing complex demand for figuring out the driving question that the students investigate throughout the unit (Krajcik & Czerniak, 2018). Each project has an overarching anchoring phenomenon and several sub-anchoring phenomena. The sub-anchoring phenomena integrate to build students' sensemaking of the overarching phenomenon. Smaller grain-size driving questions that build toward the overarching driving question are organized coherently. These smaller grain-size driving questions anchor different but related phenomena or problems to solve the project's overarching driving question. Each sub-anchoring phenomenon or problem builds off the last and offers new insight toward the driving question. This careful and purposeful building toward an understanding of the phenomena through 3D learning supports the acquisition of "usable" knowledge and builds coherence across the unit. This is what we refined and iterated over time as a critical solution to sustaining student engagement to develop knowledge-in-use (Miller & Krajcik, 2020). Please refer to Figure 5.

The selection and organization of anchoring phenomena in PBL provides enlightenment about how individuals' adaptive skills can be deliberately educated by engaging in various contexts. Based on the core idea of a learning progression (Smith et al., 2006), five or six sub-anchoring phenomena are often organized to work together to foster students' sense-making of the overarching anchoring phenomenon (project/unit phenomenon/problem). As shown in Figure 5, each sub-anchoring phenomenon interacts with the other, but it is mainly constructed based on the core ideas' learning progression. Notably, rather than decoding the case into different parts, we try to include full-view phenomena to maintain the context's authenticity and complexity. Students persist at, make sense of, and engage in real-world phenomena and cope with the complex relationships among various factors. They sustain in making sense of the phenomena through active construction, which can foster their engagement and motivation.

Figure 5. Relationships among anchoring phenomena and overarching phenomenon

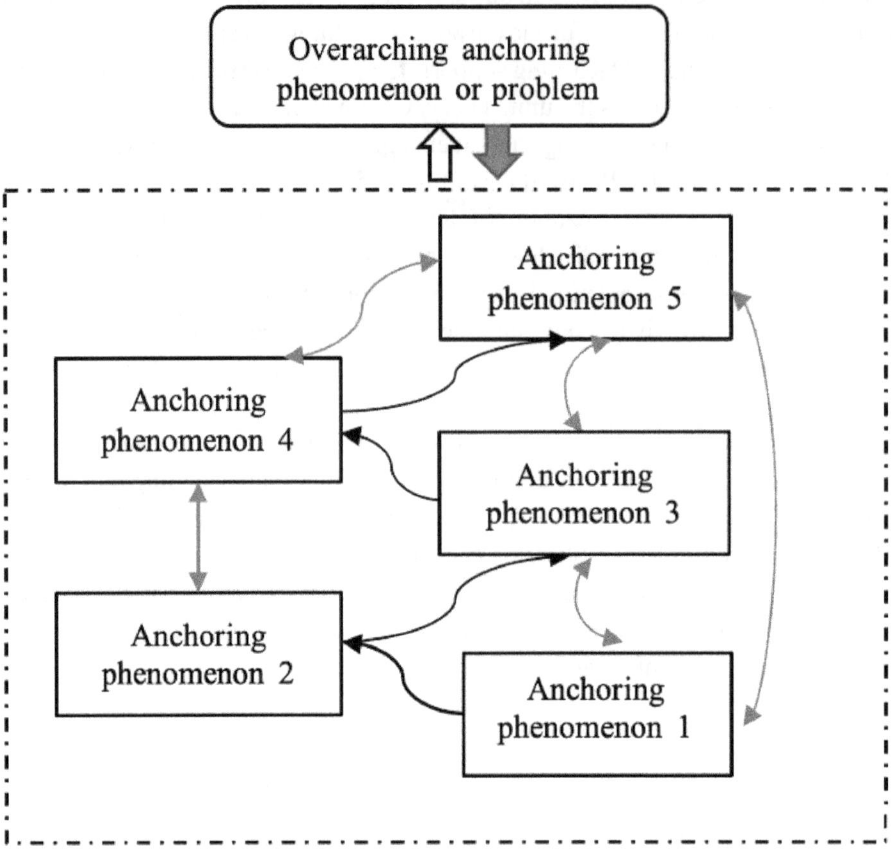

To better illustrate the relationships among the phenomena within a PBL unit, we use the above "Squirrel" unit as an example. Observing squirrels in their schoolyard opens the unit and focuses on the various phenomena and how they work together toward students' knowledge-in-use development of the concept of adaptation. Table 1 shows the organization of phenomena in the third grade "Squirrels" unit.

Table 1. Hierarchical relationships among various phenomena within a PBL unit

Level 1: Overarching phenomenon/ unit driving question Unit 3.1: Why do I see so many squirrels but I can't find any stegosauruses?	
Level 2: Learning set phenomena/ driving questions	***Level 3:*** Lesson level phenomena/ driving questions
LS1: (Squirrel survive) What do squirrels need to survive?	L1.1 What does your family know about squirrels, prehistoric organisms, and fossils?
	L1.2 What do we already know about squirrels today and in the past?
	L1.3 Do any organisms today look like those from the Jurassic period?
	L1.4 What can we observe about squirrels and how they survive outside?
	L1.5 How can I make a model to explain how a squirrel survives outside?
	L1.6 How does a squirrel survive as it grows up, escapes from predators, and lives through the winter?
	L1.7 What can we learn from each other by sharing and revising our models?
LS2: (Structure) How is the squirrel's structure unique and important?	L2.1 How does a squirrel use its teeth to survive outside?
	L2.2 How does a squirrel balance?
	L2.3 and L2.4 How can a squirrel jump so far? Math: How does a line plot help me understand how squirrels jump?
	L2.5 How do the squirrel's structures help it meet its needs and survive in its environment?
LS3: (Squirrel's environment) What other organisms live in the squirrel's environment, and does the squirrel need them to survive?	L3.1 What other organisms live in the squirrel's environment?
	L3.2 What is unique about other organisms in the squirrel's environment and how do they help the squirrel survive?
	L3.3 Are there a lot more trees in some communities than in others?
LS4: (Prehistoric organisms) How do scientists use evidence from the past and present to find out about prehistoric organisms?	L4.1 How can we use a timeline to think about the past?
	L4.2 What were past environments like?
	L4.3 What are fossils and how do they help us understand prehistoric organisms and environments?
	L4.4 Which organisms could live in prehistoric environments?
	L4.5 What do fossils tell us about which organisms were here a long time ago, which are still here, and which are new?
LS5: (Fossils as evidence) How can we use fossils to figure out how organisms change over long periods of time?	L5.1 Can we find some plants and animals outside that look like organisms from the past?
	L5.2 How can we use fossils as evidence to figure out what happened to the plants from the Jurassic?
	L5.3 How have animals from the past changed?
	L5.4 How did our prehistoric organism interact with the other organisms in its environment?
	L5.5 Math: Can we imagine how big these animals were? How would it look if they were next to each other
	L5.6: Where have scientists and other citizens found fossils from the Jurassic period? SEL focus: What do I want to know more about?
LS6: (Survival and extinction stories) Why did some animals die out and some live?	L6.1 Can we bring everything we know together to answer our driving question?
	L6.2 How can we use our model to explain our thinking about what happened to the stegosaurus?
	L6.3 How can we apply our knowledge to understand a different change to the environment?

Critical features of phenomena in a PBL unit. The overarching anchoring phenomenon and sub-anchoring phenomena interact with each other to build a coherent way of thinking. Each context in the project has three different affordances. See Figure 6 below. The overarching anchoring phenomenon should meet the criteria of engaging, authentic, challenging, and meaningful to students and have the potential to focus on some core ideas highlighted in the NGSS (2013). This process usually starts with selecting the core idea and then unpacking ideas. The smaller grain-size learning goal is generated based on unpacking the ideas in the standards, named learning performances (Krajcik, McNeil, &Reiser, 2008) developed from performance understandings by Perkins, Crismond, Simmons, & Unger, 1995. Learning performances integrate core ideas with scientific practices (Krajcik et al., 2008) and reflect working scientists' professional disciplinary practices. Another function of a learning performance is its core ideas, which guide and direct curriculum developers and teachers to come up with real-world phenomena. Even at this step, if they start with the learning performance to come up with a driving question, it is challenging to find questions that students find meaningful and interesting (Krajcik & Shin, 2014).

The overarching anchoring phenomenon and sub-anchoring phenomena should meet several essential features. These features can be divided into at least two categories, general features and features for adaptive skills (see Figure 6 below). For both overarching anchoring phenomenon and sub-anchoring phenomena that build coherently, their generic features should include: (a) a phenomenon that is authentic and engaging; (b) the phenomenon is meaningful to learners; (c) the phenomenon has an optimal level of challenge so that students will exert what they consider reasonable effort to gain success (Brophy, 1983); (d) opportunities occur to work with others (Malone & Lepper, 1987); (e) flexible opportunities exist for students to choose what and/or how the work is done; and, (f) the creation of an artifact to bring closure to the unit. Also, the overarching anchoring phenomenon should be broad enough to cover a wide range of related phenomena, and it by making sense of the phenomena students should develop knowledge of the core idea. The core idea should be based on critical ideas in the discipline or field with various associated ideas. The overarching phenomenon should also be sustainable and allow students to pursue it over time and explore ideas in detail. The sub-anchoring phenomenon should be varied and include novel elements in each phenomenon (Blumenfeld et al., 1991). Each sub-anchoring phenomenon should contribute a unique aspect to making sense of the overarching phenomenon, either in the concept aspect or context variability aspect. Key concepts build across these sub-anchoring phenomena and work together to reach the learning goal. These features enable the selection and organization of various phenomena to work integrally for developing learners' adaptive skills and enduring understanding.

Figure 6. Phenomenon features

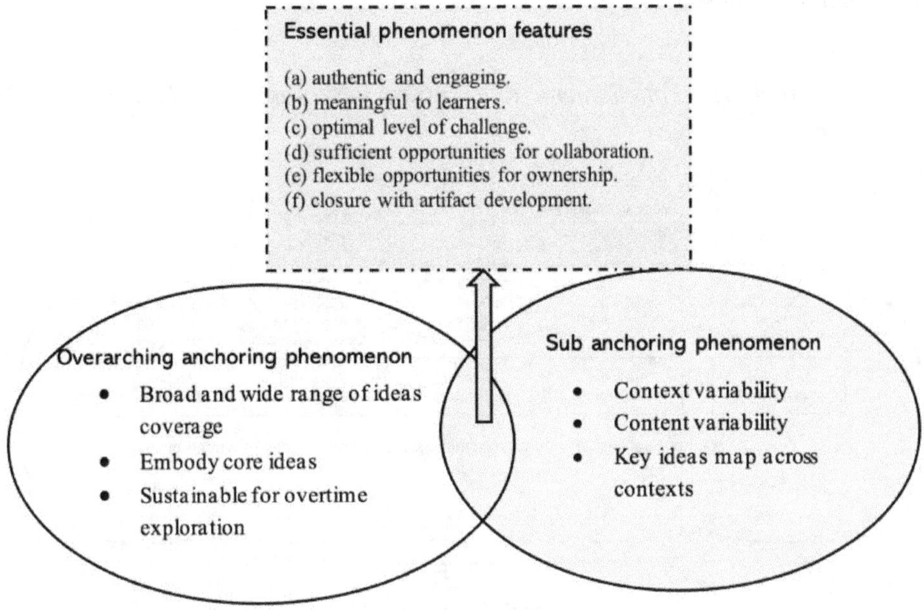

Unique affordances of each phenomenon. In terms of knowledge-in-use development, each phenomenon should have three types of affordances (see Figure 7. Below). First, content variability affordance (ontological features of adaptive worldview) provides specific content schemas. Each sub-anchoring phenomenon offers a unique aspect of the core idea of understanding, called content variability. Conceptual understanding is the primary basis for being flexible and adaptive (Ward et al., 2018). Across the sub-anchoring phenomena, students focus on various content ideas that constitute the core ideas needed to explain the overarching anchoring phenomenon and are underlined and mapped across the sub-anchoring phenomena. Second, each sub-anchoring phenomenon also affords context variability. As we discussed the phenomenon features before, each sub-anchoring phenomenon should be varied in context to provide various contextual experiences for learners. Those various contextual experiences will contribute to their adaptive decision-making when the student encounters a novel situation. The more cases in which a concept is experienced, and the more varied the cases that are experienced, the more likely it is that the concept will be learned flexibly and that adaptive skills will be developed (Rand et al., 2020). Third, there is a common scientific mindset also developed across the sub-anchoring phenomena. The common scientific mindset is one of the epistemological features of adaptive skills. In science learning, this common scientific mindset could be various scientific and engineering practices to engage learners in making sense of phenomena or solving problems.

Figure 7. Anchoring phenomena and knowledge-in-use affordances in PBL environment

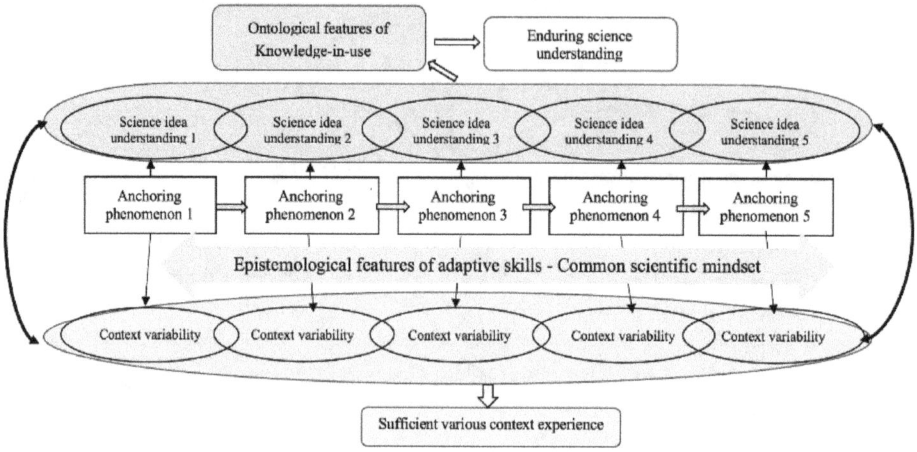

By engaging in sustainable science learning, students gain integrated knowledge. Students adaptively draw on various aspects of the phenomena they experienced to help them make sense of novel situations during the process. In doing so, learners can gain general thinking (common science mindset) and specific content schemas with various context variability, which correspond to the mechanism of CFT (Spiro, 2020).

The coherent sub-anchoring phenomena, with overall learning goals, drive students to engage in and persist at the sensemaking process. With consistent exploring, students may encounter emergent aspects of a situation, including previously unrecognized opportunities for action and new ways of achieving success. These emergent aspects could also be challenges; students must detect, make sense of, validate, and decide whether adaptive replanning is called for or not. These challenges are critical for engagement. This process also offers opportunities for learners to develop adaptive skills.

Case-Proficiency Scaling Through Collaboration With Peers and Teacher Support

In project-based learning, students participate in meaningful and engaging phenomena to make sense of the overarching phenomenon. Each sub-anchoring phenomenon is driven by a lesson-level learning performance, which is created in alignment with the unpacking of performance expectations from the Next Generation Science Standards (NGSS) (2013). The performance expectations and learning goals of each lesson are three-dimensional, allowing students to engage in the process of making sense of phenomena through various science practices such as modeling. During the project, students construct individual models, work in small groups to reach a consensus model, and eventually construct a class-consensus model. This collaborative process takes place within a supportive learning environment, where students have opportunity to engage with their peers, share their ideas with the class, and provide feedback on each other's work. The teacher provides scaffolding in their modeling and explanation construction, which is gradually faded across lessons. The diversity of learning opportunities in the project-based learning environment provides students with the chance to actively reflect on their learning and supports their development of adaptive skills.

For example, the journey of designing fun moving toys for other kids begins in a PBL unit with a driving question, "*How can we design fun moving toys that other kids can build*?" The students set out to explore this question by creating prototypes of moving toys and observing their movements to develop models that describe their pattern of motion. Throughout the unit, the students work through several learning sets, each building upon their previous knowledge and experiences.

In the first learning set, students start by observing a toy air rocket and brainstorming questions about how it moves. In small groups, they explore different types of air rockets, generate questions about how rockets work and why each moves the way it does, and share their observations and questions on the DQB. The students also introduce themselves to the driving question and share their personal experiences with moving toys.

Next, the class observes and discusses how the motion of an air rocket changes during flight. As a group, they develop, share, and revise models to represent the changes in motion. The students also view a video of a boy who comes from a very poor country and has a low SES status making his own toy. Together, they then collaborate to build simple prototypes of moving toys and record observations of the patterns in motion as the toy moves. The students continue to develop and share models that show how their toy moves in a pattern.

Finally, the students revise their models and develop engineering solutions to create their final artifact, a design portfolio that showcases how they have incorporated what they have learned throughout the unit. The journey of designing fun moving toys for other kids is an exciting and educational experience that allows students to explore their interests and knowledge while learning about patterns of motion and engineering.

Active Reflection Through Iteratively Revisiting the Driving Question Board and Creating Artifacts

Driving questions and driving question boards hold a significant impact on project-based learning. The learning process begins with a driving question, and students are encouraged to add their own related questions to a physical or electronic driving question board. They are required to regularly consult the driving questions board, both at the start and conclusion of each lesson. The students use this opportunity to summarize what they have learned, reflect on what they still need to grasp and add new questions to deepen their thinking. Additionally, students are tasked with categorizing the driving questions on the board into different groups. They label each category and allow for further clarification of their understanding. This develops students' knowledge of what they know and don't know (Ward et al., 2018). This active reflection on their learning process supports the growth of their adaptive skills.

CONCLUSION

The world we live in is changing rapidly, and real-world phenomena are complex and ill-structured. Real-world phenomena have non-linear and dynamic features that often require multiple flexible pathways to solve problems. Hence, individuals should have the ability to adapt to new situations that they encounter every day. Building on the seminal cognitive work of adaptive skills (Rand et al., 2020; Ward et al., 2018) and project-based learning (Blumenfeld et al., 1991), this chapter presented an integrated micocognitive model of adaptive skills by modifying the macrocognitive model proposed by Hoffman, Ward, and colleagues in 2014. We discussed how and why PBL has the potential to develop individuals' adaptive skills that share common features of knowledge-in-use based on the microcognitive models and adaptivity training principles. We briefly conclude our main ideas and propose several future research agendas from the two parts of microcognitive model and the development of knowledge-in-use in PBL environment.

First is the integrated microcognitive model of adaptive skills modified from the D/F+F model (Hoffman, Ward, et al., 2014). In the microcognitive model, we unfolded the latent process of adaptiveness when encountering new phenomena. Specifically, we emphasize the process of constructing the adaptive worldview across various contexts and the interaction between the adaptive worldviews and the new context to generate schema in the moment. We highlighted the importance of individuals' commonalities of, and differences in, analyzing their prior experienced contexts and the new context. This supports learners in selecting the prominent parts of each context and glean a common meta-feature of an adaptive mindset and specific content schemas from their long-term memory. The common epistemological aspect of an adaptive worldview and the specific content features are further tailored to fit the context information. The tailored epistemological adaptive worldview interacts with the tailored specific content schemas to construct a frame for the novel context. In so doing, one can make sense of the novel situation and evaluate the initial plan to adapt to their new decisions.

The microcognitive model of adaptive skills provides similar cognitive pathways of knowledge-in-use. Particularly, the model presents the idea that when developing an individual's knowledge-in-use, we should emphasize both the sensemaking process and the flexecuting process. During the processes, there is a collision between a new situation and prior knowledge and experience stored in long-term memory. This process is like a chemical reaction in which contextual variability and conceptual variability are two critical catalysts to speed up the process. With the appropriate design of the phenomena or situations, the reaction process will have different outputs and efficiencies. Therefore, to develop an individual's knowledge-in-use, we should highlight the latent developing process by considering both new phenomena and individual differences. Future research should explore the cognitive process of knowledge-in-use at a finer-grain size level by providing both theoretical underpinnings and empirical evidence. To this end, empirical evidence from a neuroscience perspective using cutting-edge innovative technology might provide valuable information. In addition, given the limitation in the capacity of an individual's working memory, future research should explore how technology or artificial intelligence can facilitate this cognitive process.

The second main part of this chapter discusses why and how a PBL environment has the potential to support an individual's knowledge-in-use development. Based upon the microcognitive model of the knowledge-in-use development process and the adaptivity training principles (Ward et al., 2018), we first presented an overview of the theoretical foundations and main design features of PBL. Then, we dove into the coherent and sustained learning environment design by highlighting the contextual variabilities and conceptual variabilities of various phenomena of a PBL unit.

In a project-based learning (PBL) environment, learners acquire a sophisticated understanding of the key ideas and practices of science that are essential for comprehending the world around them. To facilitate this knowledge acquisition, a well-designed learning environment is necessary (Fulmer, Tanas, & Weiss, 2018; Miller & Krajcik, 2019). This perspective aligns with previous research in the field of situated learning (Lave, 1988). Lave and colleagues (1991) emphasized the social and interactive nature of learning, proposing that it typically occurs in "communities of practice." Greeno and colleagues (1993) offered a reconceptualization of transfer based on the affordances (Gibson, 1977) of different situations and contexts. Their perspective highlights the potential for transfer to occur when similar affordances are present across different situations.

While this is just a preliminary discussion, there are several areas that require further exploration. These areas include the role of motivation and engagement in designing productive learning environments (Blumenfeld et al., 1991). Additionally, future research may explore how the PBL environment can be designed to promote both motivation and cognitive engagement (Mayer, 2014). Another important aspect is designing PBL environments with an equity and social-justice orientation, taking into account the diverse cultural backgrounds and individual needs of students to better support their knowledge development (Adah Miller, et al., 2022). Moreover, the role of teachers in implementing PBL is critical. Teachers are instrumental in shaping the learning environment and guiding students in their acquisition of science knowledge (Li et al., 2021). We call for future studies to investigate how to better understand and enhance the role of teachers in the PBL setting.

REFERENCES

Adah Miller, E., Codere, S., Charara, J., Severson, A., & Spaul, H. (2022). Fire Ecology in Fourth Grade: Moving Place-based Education to the Global Stage. Suleiman, M., & Huber, T. (eds.) Beyond Provincialism: Promoting Global Competencies in Teacher and Educator Preparation. Information Age Publishing.

Adah Miller, E., Makori, H., Akgun, S., Miller, C., Li, T., & Codere, S. (2022). Including teachers in the social justice equation of project-based learning: A response to Lee & Grapin. *Journal of Research in Science Teaching*, *59*(9), 1726–1732. doi:10.1002/tea.21805

Anderson, J., & Li, Y. (Eds.). (2020). *Integrated approaches to STEM education: An international perspective*. Springer Nature. doi:10.1007/978-3-030-52229-2

Beneke, S., & Ostrosky, M. M. (2009). Teachers' Views of the Efficacy of Incorporating the Project Approach into Classroom Practice with Diverse Learners. *Early Childhood Research & Practice*, *11*(1), n1.

Blumenfeld, P. C., Soloway, E., Marx, R. W., Krajcik, J. S., Guzdial, M., & Palincsar, A. (1991). Motivating project-based learning: Sustaining the doing, supporting the learning. *Educational Psychologist, 26*(3-4), 369–398. doi:10.1080/00461520.1991.9653139

Bransford, J. D., Brown, A. L., & Cocking, R. R. (2000). *How people learn* (Vol. 11). National academy press.

Charara, J., Miller, E. A., & Krajcik, J. (2021). Knowledge in Use: Designing for Play in Kindergarten Science Contexts. *Journal of Leadership, Equity, and Research, 7*(1).

Chu, S. K. W., Tse, S. K., & Chow, K. (2011). Using collaborative teaching and inquiry project-based learning to help primary school students develop information literacy and information skills. *Library & Information Science Research, 33*(2), 132–143.

Cobb, P., & Bowers, J. (1999). Cognitive and situated learning perspectives in theory and practice. *Educational Researcher, 28*(2), 4–15. doi:10.3102/0013189X028002004

Efstratia, D. (2014). Experiential education through project based learning. *Procedia: Social and Behavioral Sciences, 152*, 1256–1260. doi:10.1016/j.sbspro.2014.09.362

Esposito, A. G., & Bauer, P. J. (2017). Going beyond the lesson: Self-generating new factual knowledge in the classroom. *Journal of Experimental Child Psychology, 153*, 110–125. doi:10.1016/j.jecp.2016.09.003 PMID:27728784

Finnish National Board of Education. (2016). *National Core Curriculum for Basic Education 2014*. Helsinki, Finland: Finnish National Board of Education. .

Fulmer, G. W., Tanas, J., & Weiss, K. A. (2018). The challenges of alignment for the Next Generation Science Standards. *Journal of Research in Science Teaching, 55*(7), 1076–1100. doi:10.1002/tea.21481

Geier, R., Blumenfeld, P. C., Marx, R. W., Krajcik, J. S., Fishman, B., Soloway, E., & Clay-Chambers, J. (2008). Standardized test outcomes for students engaged in inquiry-based science curricula in the context of urban reform. *Journal of Research in Science Teaching, 45*(8), 922–939. doi:10.1002/tea.20248

Harris, C. J., Penuel, W. R., D'Angelo, C. M., DeBarger, A. H., Gallagher, L. P., Kennedy, C. A., Cheng, B. H., & Krajcik, J. S. (2015). Impact of project-based curriculum materials on student learning in science: Results of a randomized controlled trial. *Journal of Research in Science Teaching, 52*(10), 1362–1385. doi:10.1002/tea.21263

He, P., Chen, I.-C., Touitou, I., Bartz, K., Schneider, B., & Krajcik, J. (2023). Predicting student science achievement using post-unit assessment performances in a coherent high school chemistry project-based learning system. *Journal of Research in Science Teaching*, *60*(4), 724–760. doi:10.1002/tea.21815

Helle, L., Tynjälä, P., & Olkinuora, E. (2006). Project-based learning in post-secondary education–theory, practice and rubber sling shots. *Higher Education*, *51*(2), 287–314. doi:10.100710734-004-6386-5

Inkinen, J., Klager, C., Juuti, K., Schneider, B., Salmela-Aro, K., Krajcik, J., & Lavonen, J. (2020). High school students' situational engagement associated with scientific practices in designed science learning situations. *Science Education*, *104*(4), 667–692. doi:10.1002ce.21570

Krajcik, J., McNeill, K. L., & Reiser, B. J. (2008). Learning-goals-driven design model: Developing curriculum materials that align with national standards and incorporate project-based pedagogy. *Science Education*, *92*(1), 1–32. doi:10.1002ce.20240

Krajcik, J. S., Blumenfeld, P. C., Marx, R. W., & Soloway, E. (1994). A collaborative model for helping middle grade science teachers learn project-based instruction. *The Elementary School Journal*, *94*(5), 483–497. doi:10.1086/461779

Krajcik, J. S., & Czerniak, C. M. (2018). *Teaching science in elementary and middle school: A project-based learning approach*. Routledge. doi:10.4324/9781315205014

Krajcik, J. S., Palincsar, A., & Miller, E. (2015). *Multiple literacy in project-based learning*. Michigan State University.

Krajcik, J. S., Schneider, B., Miller, E., Chen, I. C., Bradford, L., Peek-Brown, D., Bartz, K., Baker, Q., Palicsar, A., Miller, C., & Codere, S. (2023). Assessing the Effect of Project-Based Learning on Science Learning in Elementary Schools. *American Educational Research Journal*, *60*(1), 70–102. doi:10.3102/00028312221129247

Krajcik, J. S., & Shin, N. (2014). Project-based learning. In R. K. Sawyer (Ed.), The Cambridge handbook of learning sciences, (2nd ed.), New York: Cambridge. doi:10.1017/CBO9781139519526.018

Krajcik, J. S., & Shin, N. (2023). *Student Conceptions*. Conceptual Change, and Learning Progressions.

Kulgemeyer, C., & Schecker, H. (2014). Research on Educational Standards in German Science Education - Towards a model of student competences. *Eurasia Journal of Mathematics, Science and Technology Education*, *10*(4), 257–269. doi:10.12973/eurasia.2014.1081a

Lee, O., & Grapin, S. E. (2022). The role of phenomena and problems in science and STEM education: Traditional, contemporary, and future approaches. *Journal of Research in Science Teaching, 59*(7), 1301–1309. doi:10.1002/tea.21776

Li, T., Chen, I., Miller, E., Miller, C., Schneider, B., & Krajcik, J. (Accepted). The Relationships between Elementary Students' Knowledge-In-Use Performance and Their Science Achievement. *Journal of Research in Science Teaching.*

Li, T., Miller, E., Chen, I.C., Bartz, K., Codere, S., & Krajcik, J. (2021). The relationship between teacher's support of literacy development and elementary students' modelling proficiency in project-based learning science classrooms. *Education 3-13: International Journal of Primary, Elementary and Early Years Education.* . [ESCI]. doi:10.1080/03004279.2020.1854959

Miller, E., & Reigh, E. (2020). *Interactional approach to language: Supporting all students to engage in equitable sensemaking.*

Miller, E. C., & Krajcik, J. S. (2019). Promoting deep learning through project-based learning: A design problem. *Disciplinary and Interdisciplinary Science Education Research, 1*(1), 1–10. doi:10.118643031-019-0009-6

Ministry of Education, (People's Republic of China). (2003). National standards of general senior secondary school chemistry curriculum. *People's Education Press.* (in Chinese)

Ministry of Science and Technology and Publicity Department. (2016). *The Chinese Citizens Science Quality Benchmark.* Ministry of Science and Technology and the Publicity Department. http://www. gov.cn/gongbao/content/2016/content_5103155.htm

National Academies of Sciences, Engineering, and Medicine. (2019). *Science and Engineering for Grades 6-12: Investigation and Design at the Center.* Washington, DC: The National Academies Press.

National Research Council. (2011). *Assessing 21st century skills: Summary of a workshop.* The National Academies Press.

National Research Council. (2012a). *A Framework for K-12 science education: practices, crosscutting concepts, and core ideas.* The National Academies Press.

National Research Council. (2012b). *Education for life and work: Developing transferable knowledge and skills in the 21st century.* National Academies Press.

Next Generation Science Standards Lead States. (2013). *Next generation science standards: For states, by states.* Next Generation Science Standards.

OECD. (2016). PISA 2015. In *Assessment and analytical framework: Science, reading, mathematic and financial literacy*. OECD Publishing.

Oyewo, O. A., Ramaila, S., & Mavuru, L. (2022). Harnessing Project-Based Learning to Enhance STEM Students' Critical Thinking Skills Using Water Treatment Activity. *Education Sciences*, *12*(11), 780. doi:10.3390/educsci12110780

Pellegrino, J. W., & Hilton, M. L. (2012). *Committee on defining deeper learning and 21st century skills*. Center for Education.

Penuel, W. R., Reiser, B. J., McGill, T. A., Novak, M., Van Horne, K., & Orwig, A. (2022). Connecting student interests and questions with science learning goals through project-based storylines. *Disciplinary and Interdisciplinary Science Education Research*, *4*(1), 1–27. doi:10.118643031-021-00040-z

Potvin, A. S., Miller, E. A., Kuck, R., Berland, L. K., Boardman, A. G., Kavanagh, S. S., Clark, T. L., & Cheng, B. H. (2022). Mapping Enabling Conditions for High-Quality PBL: A Collaboratory Approach. *Education Sciences*, *12*(3), 222. doi:10.3390/educsci12030222

Reiser, B. J. (2014). *Designing coherent storylines aligned with NGSS for the K-12 classroom*. Boston: National Science Education Leadership Association http://bit.ly/1TZeSaG

Schneider, B., Krajcik, J., Lavonen, J., & Salmela-Aro, K. (2020). *Learning Science: The Value of Crafting Engagement in Science Environments*. Yale University Press.

Schneider, B., Krajcik, J., Lavonen, J., Salmela-Aro, K., Broda, M., Spicer, J., Bruner, J., Moeller, J., Linnansaari, J., Juuti, K., & Viljaranta, J. (2016). Investigating optimal learning moments in US and Finnish science classes. *Journal of Research in Science Teaching*, *53*(3), 400–421. doi:10.1002/tea.21306

Schneider, B., Krajcik, J., Lavonen, J., Salmela-Aro, K., Klager, C., Bradford, L., Chen, I.-C., Baker, Q., Touitou, I., Peek-Brown, D., Dezendorf, R. M., Maestrales, S., & Bartz, K. (2022). Improving science achievement—Is it possible? Evaluating the efficacy of a high school chemistry and physics project-based learning intervention. *Educational Researcher*, *51*(2), 109–121. doi:10.3102/0013189X211067742

Shin, N., Choi, S.-Y., Stevens, S. Y., & Krajcik, J. S. (2019). The impact of using coherent curriculum on students' understanding of core ideas in chemistry. *International Journal of Science and Mathematics Education*, *17*(2), 295–315. doi:10.100710763-017-9861-z

Tal, R. T., & Miedijensky, S. (2005). A model of alternative embedded assessment in a pull-out enrichment program for the gifted. *Gifted Education International*, *20*(2), 166–186. doi:10.1177/026142940502000208

Thomas, J. W. (2000). *A review of research on project-based learning*. Autodesk Foundation.

Zhao, Y., & Wang, L. (2022). A case study of student development across project-based learning units in middle school chemistry. *Disciplinary and Interdisciplinary Science Education Research*, *4*(1), 1–20.

ENDNOTE

1 Open and complete unit can be found from this link: https://sprocket.lucasedresearch.org/course/science3/squirrels

Chapter 2
Progressive Pedagogy for Extended Reality:
Shaping and Facilitating the Fourth Industrial Revolution

Catherine Hayes
University of Sunderland, UK

ABSTRACT

This chapter provides an insight into the theoretical perspectives which form the foundation of extended reality (XR) and its emergence in practice as a fundamental part of science curricula. Issues such as the authenticity of learning, the validity and reliability of XR within processes of assessment, and the theoretical underpinnings of pedagogical approaches in health professions pedagogy are illuminated. Also considered are the implications of XR within the context of science-based learning and the delineation of cognitive, affective, and psychomotor domains of learning in relation to patient outcomes at the front line of care in applied scientific practice. The Covid-19 pandemic, which has impacted all science based pedagogic practice since March 2020, is also considered in the context of moves to ensure that science education can continue, albeit via hybrid models of learning as opposed to traditional pedagogical approaches which have remained little altered over the last century.

DOI: 10.4018/978-1-6684-6932-3.ch002

INTRODUCTION

"Intelligence is the ability to adapt to change..." Stephen Hawking (1942-2018)

The justification of pedagogy in the context of Extended Reality (XR), which encompasses Virtual Reality (VR), Augmented Reality (AR) and Mixed Reality (MR), has become an ongoing source of complex ambiguity over the last decade, which the COVID-19 pandemic has exacerbated (van der Niet and Bleakley, 2021). Ensuring the validity and reliability of XR experiences within science education remains central to the potential to rule out technologies as adjuncts to optimal pedagogic practice as an authentic means of providing insight and illumination of scientific contexts, scenarios and processes (McGrath et al., 2018). For the purposes of this chapter there will be four fundamental operationally definitive terms of what the umbrella term XR encompasses. Firstly VR refers to the use of computer technology in the creation of simulated learning environments. Secondly, AR pertains to the addition of computerised content as an overlay to reality, which means that learners can actively interact both with real world and augmentations of it at the same time. Mixed or hybrid reality refers to the transection of virtual worlds and actual worlds, where physical and computerised objects can interact and exist concurrently. XR encompasses all of these and as a collective group they have revolutionised science and technological training, particularly in relation to the practise of risk management and professional role identity in life and death medical situations such as obstetric emergencies, as reported by Hayes, Hinshaw and Petrie (2018).

Training for the strategic management of risk in science-based healthcare practice in situated contexts of healthcare provision has been a key focus in the use of XR in practice (Hilty et al, 2020). Not only does it involve rational aspects of cognitive knowledge or the absolute adherence to demonstration of psychomotor skills and affective domain learning (Zulkilfli, 2019). It also encompasses the intuitive, tacit and largely intangible intellectual instincts that develop with sustained experiential learning (Humpherys, Bakir and Babb, 2021). One of the key issues has been the challenge of pedagogical approaches in practice. XR has enabled the benchmarking of perceived levels of interprofessional and multidisciplinary teamwork, where intuitive knowledge can be used to measure risk, regardless of the level of the organisational hierarchy within which personnel are employed (Goh and Sandars, 2020; Hayes, Hinshaw and Petrie, 2018). This chapter explores key epistemological perspectives that can be used to ensure the level of authenticity required to highlight the pedagogical shifts in the application of learning theory that now characterise responsive scientific curriculum design and adaptation to accommodate XR in practice.

THE CURRENT CONDITION OF SCIENCE PEDAGOGY

The ongoing pandemic, which by October 18th, 2022, had reached 625 million confirmed cases and 6.25 million fatalities, has not only changed the world of science education in its current form, but has also presumably altered its mechanism of delivery for the foreseeable future (WHO International Data, 2021). In practice, the pandemic led to university closures, a switch to a hybrid learning model of education, and reduced people's ability to engage in face-to-face meetings (Okoye et al, 2021). The plethora of academic articles surrounding online learning is phenomenal, but few actually address how a fundamental paradigm shift in Higher Education is implemented methodologically (Luctkar-Flude and Tyerman, 2021). What is usually described as a narrative description of the pragmatic use of technological intervention, rather than any degree of alignment in terms of underpinning theoretical justification for implementation, or constructive alignment demonstrating how assessment processes can be effectively driven by complementary processes of teaching and learning (Moreira, 2020). The physicality of learning has also been altered beyond recognition, seeing people as upper torsos and faces has changed everyday interaction in the situated nature and context of learning, yet minimal evidence exists as to the long-term impact of this on motivating, engaging, and sustaining active processes of learning, teaching and assessment on an individual level (Park and Kim, 2020). By over reliance on the physicalism of the articulated voice and postural positions of the upper torso, executed in an atmosphere of scrutiny, learners have had to change their interaction so that their degree of interaction is heavily influenced and constrained (Obrad, 2020). Whilst predicting how global disease and inequality will influence the future of education, it is impossible to ignore potential challenges that lie ahead for science education (Bevins et al, 2020). It is possible to inferentially predict that COVID-19 may be one of the first of a new generation of global pandemics that will need to be dealt with, alongside dramatic changes in overall global warming, which will also ensure that populations which are densely populated suffer most severely (Negev et al, 2021). Alongside the issue of pandemic disease is the prospect of global catastrophes and natural disasters occurring far more frequently and necessitating support and solution now. The progressive redevelopment of existing science and technology to accommodate this is more than apparent, so that learners can engage physically in scientific settings with smaller equipment and a greater capacity to seamlessly integrate extended reality into all practice (Yigitcanlar et al, 2020). Geographically, there are other issues at play in terms accessibility to technology on a global scale, because of which some countries cannot be guaranteed adequate internet access, bandwidth or the cost of the technology products may simply be prohibitively expensive (Horton, 2021). Being cognisant of this necessitates ensuring both affordability and accessibility across the globe if the differential inequity between countries is to be addressed on an ongoing basis. Being able to standardise and regulate experience for science learners is also of fundamental importance if equity and parity of experience are to be assured across these programmes (Crouch et al, 2021).

CURRICULUM JUSTIFICATION, DESIGN AND DEVELOPMENT OF EXTENDED REALITY IN HEALTHCARE EDUCATION

Pedagogical design and its address within the context of science education curricula often places processes of curriculum justification and design under scrutiny (Annala et al, 2021). This is largely attributable to the complex ambiguity that curriculum designers have to contend with, in terms of the technical capacity of XR equipment and the accompanying resources they necessitate (Aebersold and Dunbar, 2021). A key example of this is the multidisciplinary perspectives that designed scenarios must authentically represent within the context of science education, where XR is implemented in practice (Antoniou et al, 2020). Because of this, the concept of experiential learning has become a focus for the opportunity of ensuring that XR is relevant to real world application in practice, with the introduction of its continuum's integral parts (AR, VR and MR). The published evidence to date demonstrates that XR training has been shown to improve learner performance skills across an array of learning domains (cognitive, affective and psychomotor) within the context of instructional simulation (Tabatabei, 2020). Having moved firmly from proof-of-concept pilot studies, XR as an embedded part of health and medical curricula is being used across a wide variety of simulated contexts (Tang et al, 2020). In comparison to traditional didactic teaching methodologies the benefits are seen within the areas of student motivation, confidence and the capacity of learners to take measured risk within the context of a consequence free facilitative environment where constructive feedback on performance can inform reflexive praxis around complex ambiguity and clinical decision making. Acclimatisation to new contexts and settings, for military learners, for example is invaluable, where introduction to new unfamiliar climates and contexts is undertaken. This is particularly significant where interactive immersive scenarios are used as experiential learning opportunities. (Cobos, 2020). The addition of iteratively increasing levels of complex ambiguity as skill increases is another opportunity for the formal scaffolding of learning within customised immersive experiences (Orr et al, 2020). Depending on whether learning is about the magnification of micro-theoretical concepts like atomic level particles, blood corpuscles or synaptic junctions or whether learning is centred around macro social constructs such as social justice and equity in different contexts of culture or climate, immersive technology with XR has the potential to extend the reach within and between academic pedagogies and disciplines; and as a direct consequence impact on the capacity and capability of multidisciplinary and interdisciplinary professionals working at the front line of science, regardless of context (Mitchell and Boyle, 2021). Confidence is a significant factor that underpins science learners' motivation and elevates levels of competence, ensuring this iterative cycle continues (Owens, 2021).

Roussin and Weinstock (2017) detailed the pedagogical challenges educators and trainers face leading science-based simulation training in the context of healthcare and medical services. Regarding the theoretical underpinning of their work, they presented the complexities of managing science programmes in relation to single and double-loop experiential learning and the impact of organising organisational hierarchies on the need to gain experiential learning in practice. Whilst their work omits specific reference to the implicit value of tacit knowledge in crises, the functional insight they provided about situated learning is invaluable compared to the overall administration of optimal multidisciplinary and interprofessional teamwork scenarios. Most significantly, their research demonstrated the application of scaffolded knowledge to the point of autonomy. This is consistent with Vygotsky's Zone of Proximal Development (1978) but also Argyris' acknowledgement of the relevance of single and double-loop learning. Both raise important issues in terms of the longitudinal provision in science education with XR in relation to the phasing and iterative presentation of learning outcomes, each serving to consolidate consequent stages of learning within and between cognitive, affective and psychomotor learning domains.

STRATEGICALLY DRIVING AUTHENTIC SCIENTIFIC LEARNING

Pedagogical design of blurring where reality and actuality co-exist within the context of digital immersion underpins the concept of validity in XR. While pedagogical research and applied praxis are still at the mercy of technology to a certain extent, as the virtual world becomes progressively more advanced, so will the capacity and capability for authenticity and validity within science education and training (Ligtart et al., 2021).

It is useful to consider the degree to which VR, AR and XR have been integrated across science professions within differing signature pedagogies, as well as the cognitive, psychomotor, and affective learning domain skills that must be addressed. It is here that some scientific medical and healthcare professions can be delineated in terms of their functional involvement in patient-centred care. For example, dentistry and physiotherapy have relatively practical interventions, whilst the work of a podiatrist, due to the ergonomic positioning of a practitioner, entails more opportunities to engage psychosocially with patients during their appointments (King et al., 2018). Indeed, academic debates have historically been posited about the functional basis of such healthcare professional roles as dentistry since they are so practical that people questioned whether they could be regarded as a profession (Welie, 2004). The other implication is the stage of learning that medical and allied healthcare practitioners have reached in their career trajectories, where VR, AR and XR may be implemented most appropriately. Most recently published literature focuses on learning 'in situ'. In contrast, the assessment of learning longitudinally in the context of real-life scientific praxis is often evaluated with a degree of tokenism by those whose initiatives have introduced simulation in the first place.

The operationally definitive terms of VR, AR and XR are significant when considering the nature of the artificial environments created in the exploration of potentially risk-filled scenarios, which emulate the real scientific world (Alnagrat, Ismail and Idrus, 2021). This is primarily about how artificial intelligence can be embedded within the context of virtual and simulated learning experiences, which also must be optimally facilitated, for example, by staff skilled in the field of clinical simulation (Abbas, Kenth and Bruce, 2020). Central to progression in these aspects of a pedagogical paradigm shift is the need for active dissemination and sharing of knowledge creation and acquisition that happens at the front line of scientific medical and healthcare education where simulation in the context of XR is the expected and anticipated norms of science education (Luo et al., 2021).

IMMERSIVE LEARNING EXPERIENCE WITH EXTENDED REALITY

AR and VR enhance science education by immersing students in learning experiences, positively influencing levels of engagement and motivation and by creating an atmosphere of interactive dialogue, which in terms of social inclusion, can have a direct impact on reducing learners' stress and anxiety (Brandon, Freiwirth and Hjersman, 2021). Encompassed by the umbrella term of extended reality, AR, MR and VR all bring a signature technological approach to applied praxis in the context of healthcare professions. In instances where different levels of sensorial experiences are now widely used to extend the reach of pedagogic practice in medical and healthcare education and training, this has broad ramifications not only on the practical delivery of emergent technology in scientific practice, but also on their pedagogical theoretical justifications (Robert, 2021). In instances where the affordability and hence the extent of digital technology available determines accessibility in practice-based settings. In terms of a hierarchy of availability, AR is perhaps the most widely known and least invasive of all XR and can be easily accommodated by readily available display systems (Suryanti et al., 2020).

In instances where the need for the relevance of space, situatedness and context in scientific inquiry is of fundamental importance, then MR is the best platform for the integration of experiential learning into available physical environments. In this sense, MR fills the gap that VR leaves in permitting users to interact physically with pre-identified physical spaces (Silén et al., 2008). Dimensionality is also of key importance in terms of whether the ability to facilitate visualisation in 3-D is something which can actively contribute to the acquisition of tacit knowledge and actively contribute to a more finely tuned intuitive response to any given scenario in scientific medical and healthcare practice (Gerup, Soerensen and Dieckmann, 2020). MR effectively bridges the gap between AR and VR due to advancements in display screens and visualisation resources (Juraschek, 2018).

Research into XR pedagogical initiatives and methodological designs is usually characterised by relatively small-scale case studies, focused on the collection of qualitative data rather than tangible quantitative evidence about the impact of XR in practice (Hamilton, 2021). Those which have succeeded in this, tend to focus specifically on psychomotor skill or level of cognitive improvement rather than affective domain learning. The concept of tacit knowledge is also one which remains relatively underexplored, and which remains almost a hidden bonus in developing the confidence of collective team hierarchies in clinical and medical practice. making generalisability of findings to other contexts and settings. Whilst this is a disadvantage, the potential for transferability of findings within and between similar situational contexts and settings across the global can be quite high in terms of overall content validity (Levitt, 2021).

The majority of XR technologies now extends the bridge between simulated scientific environments and their polar opposites in physical reality. The rate of progressive developmental platforms has been amplified and magnified by the global COVID-19 pandemic, which has contributed to a paradigmatic shift in learner centred hybrid and hyflex learning and teaching opportunities across the science professions pedagogies (Acharya, 2021). As signature disciplines, scientific professions such as medicine have maintained a certain degree of reticence to engage fully immersive technology within the context of clinical healthcare practice, since this completely detaches the clinician from their normal parameters of visual capacity. Whilst lenses are acceptable, they also permit and maintain human characteristics and gestures such as established eye contact and that degree of connection, which patients deserve and need (Jongerius et al, 2020). This has been particularly evidenced within the context of brain and cardiac surgery, both of which have optimally utilised these scientific and technological advances for the patients they serve.

TRADITIONS IN SCIENCE BASED MEDICAL AND HEALTHCARE EDUCATION

Across the continuum of signature pedagogies that contribute directly to science based medical and allied health education, it is anatomy and mathematics which arguably transcend most disciplinary boundaries (Hayes and Capper, 2020). The visualisation techniques, which both permit and facilitate enhanced pedagogical teaching methodologies respectively appeared first in the context of scientific simulation within the aviation industry at the beginning of the 20th century. Their developmental progression over the last century and the early part of the 21st century has ensured that levels of complex sensory and visual experience have now become a mainstay of medical and healthcare education and training programmes (Jentsch and Curtis, 2017). In the sense of providing students with the opportunity to undertake complex and risk inclined procedures, simulation is an invaluable resource. Within the context of safety, processes of experiential learning from practice can be fully exploited without there being any tangible, negative consequences in reality, so that unwanted outcomes can be illustrated and acknowledged without them actually ever having to come to fruition to do so.

Resources and student access to them have been a key issue during the pandemic, particularly for health and medical students for whom access to real- life scientific anatomical models and cadavers may have been completely negated (Hilburg et al, 2020). For these students XR has ensured an opportunity for continued engagement with the discipline of anatomy and physiology across a range of contexts and settings. Since learning is extensive in relation to the need to develop a 3-D knowledge of bodily organs and systems, then the spatiality that XR offers also permits dynamic interaction, which otherwise would not be possible. In relation to the functional role of bodily organs and systems, XR has the added advantage that simulation of the dynamic motion of the body can be visualised as a learning resource, in contrast to cadavers, which are obviously non-functional (Ziker, Truman and Dodds, 2021). Within this simulation, it is also possible to identify, isolate and examine the individual functionality that specified aspects of these systems perform via repeated familiarisation. Ensuring science learners iteratively progress from novice to expert within the use of XR of this nature, also ensures that learners can work in a risk free environment, which nevertheless serves to introduce them to the experiential learning which can form a solid foundation for their future professional careers in healthcare and medicine. In terms of pedagogic design and academic rigour the use of XR also provides an ideal opportunity to benchmark functional skills acquisition, alongside the real-time assessment of clinical practice and the opportunity for them to be exposed to sensory feedback from the situated learning they undertake (Kang et al, 2021).

BUILDING CAPABILITY FOR CLINICAL EDUCATORS WITH XR

The education and training of future scientific, medical and healthcare staff is not only dependent on disciplinary expertise but also the capacity to provide optimal learning experiences which in turn drive processes of assessment which are authentic and have real world validity and credibility (Voštinár et al, 2021). Future practitioners of all disciplines across the health and wellbeing continuum not only have to use evidence-based practice to discern clinical decision making, they also have to continually use scientific published evidence to iteratively update their continuing professional development in practice (Agha, 2021). As methodologies of pedagogic practice have become progressively more, there has been increasing emphasis on a moving away from traditional rote learning to mechanisms of interactive and engaging processes of learning, which ensure deep rather than superficial learning across all aspects of education and training. Alongside this, is the recognition that learning driven purely by pressure to 'perform' within the context of examinations is futile in the truest sense of meaningful learning opportunity. To achieve this, XR can be regarded as a significant tool in the armoury of scientific, medical and healthcare educators who use this form of specialist resource to reinforce, motivate and enhance learning opportunities through authentically, strategically and constructively aligned academic curricula in Higher Education (Howell and Mikeska, 2021). Future practitioners also need the capacity to be educated in authentic situated learning contexts which are:

- Based on principles of diversity, equity and inclusion in terms of both optimal provision and optimal service to global, national, regional and local practice areas.
- Built on an ethos of compassion for all recipients and a respect for human life and wellbeing.
- Co-constructed with medical and healthcare service users and their families and carers.
- Designed on the basis of evidence-based healthcare practice and evidence-based processes of educational provision.
- Drivers of high quality multidisciplinary and interprofessional patient centred teamworking.
- Effective in their capacity to drive education in relation to health and medical outcomes for patients at the front line of clinical care.
- Globally responsive and globally focussed
- Pedagogically designed and responsively designed on the basis of need and healthcare priorities in applied practice.
- Responsive in their capacity to dynamic and iterative change in healthcare landscapes.

- Specifically designed to value and promote autonomous learning, capacity for teamwork and proactive higher order critical thinking in relation to problem solving in health and medical care practice.

Within the context of XR integration, each pose individual challenges in relation to ensuring that digital technology is only ever seen as an adjunct to scientific, medical and healthcare education for the human race, not merely a functionalist or solely objective driver of it (Gandolfi, Kosko and Ferdig, 2021).

EXPERIENTIAL LEARNING THEORIES

The grander philosophical debates of perceived versus actual reality on an individual and collective basis are central to the processes of science teaching and learning with XR. As a tool for the enhancement of understanding, meaning making and then critical reflection on practice, digital technology in invaluable as an adjunct to the human facilitation of scientific knowledge, skills, attitudes and behaviours (Logeswaran et al 2021). The philosophical underpinning of experiential learning has its origins in constructivism, which is central to the acknowledgement of truth, either perceived or actual, and the capacity of all individuals to construct a perspective on truth based on their immersion in experience and the meaning they make of it (Koufidis et al, 2021). In this sense education is an entirely different experience for everyone who experiences the same phenomena, and it is this individual experience and perspective meaning making, which lies at the heart of both collective and personal processes of transformation. The construction of a version of reality ensures that different perspective lenses can be applied to the same phenomena, so that scrutiny from a number of sources of triangulation can take place. Whether the experience differs collectively between adults and children in terms of capacity for meaning making through experiential learning was the source of seminal debate in the late 20th century, with Knowles assertion that there was a distinction between andragogy, in terms of 'adult' ways of knowing and that of 'pedagogic' or child-based learning. What is arguable more relevant is the concept of lifelong learning, across which learning takes place from the cradle to the grave, as a temporal mechanism of integrating the prospective with the retrospective and making inference from it as part of the integral processes of reflection and reflexivity (Loeng, 2021).

THEORETICAL UNDERPINNINGS OF EXTENDED REALITY PEDAGOGY

Experiential learning has become a byword for the placement experiences that scientific, medical and healthcare students undertake as an integral part of their undergraduate studies with patients, whose consent, briskly acquired provides novice learners with an opportunity to build confidence and be guided by an ever vigilant and highly respected scientific, health or medical disciplinary expert observer. In instances of postgraduate scientific study, it may conjure the image of a professional moving from competence to proficiency and consequent mastery of their specific field of scientific disciplinary context (Mortimore et al 2021). Medical training to the uninitiated is instilled with the imagery of cadaveric dissection as a means of visualising what lies beneath the skin that is to be negotiated across a respected lifelong career. In contrast to practice-based experience at the front line of patient care in the context of care provision, though, the two are very different. It is here that whilst disciplinary knowledge and skill are invaluable to the expert educator, that skills of pedagogical praxis are equally important if the acquisition of knowledge which will underpin this lifelong career is to be assured and validated in practice. The theoretical underpinnings of learning and teaching have resonance in relation to the eventual capacity of medical and healthcare students not only to make meaning of experience but also to question that which cannot be taught, the tacit and reflexive response to intuition, which can only be derived from inner conscience and derived from processes of experiential learning in the psychomotor, cognitive and affective domains of learning, which can be tangible assessed and used to benchmark fundamental knowledge and skills acquisition (Birt et al, 2018). Similarly the capacity to transfer knowledge from the context of learning to the real world with patients who have an immediate need for assessment, diagnosis and consequent management is highly relevant to the potential to scaffold knowledge as scientific learners move from the state of disciplinary novice to expert (Durning & Artino 2011)

MEANING MAKING IN SCIENTIFIC HEALTHCARE PRACTICE

At the heart of capacity to learn and build on pre-existing experience is the theory of constructionist philosophy which when integrated alongside sociocultural perspectives from processes of active participation lies at the heart of all learning. The delineation between and linkage across theory and practice has become a fundamental goal of all involved in workplace education and training (Dennick, 2016). Since learning is both an individual and collective experience in the same sense as learning. What the

Covid-19 pandemic detracted from was the fact that education is a social science, wholly reliant on interactivity and the derivation of meaning from multi-faceted and complex situational specificities, which when grouped as a collective constitute human experience. Being able to understand and frame the margin between knowledge and scientific research entails a consideration of both the origin and consequent history of learning (Ten Cate and Billet, 2014). Alongside this, it is also necessary to consider how knowledge acquisition is gradually scaffolded through experience from novice to expert and to value those contributors to pedagogic practice who have acknowledged and formally recognised this (Graf et al, 2020). Whilst Kolb is often regarded as the seminal theorist of experiential learning, however there are several others of direct relevance to the pedagogical implementation of XR (Heong et al, 2020). Rooted in constructivist perspectives, Kolb's is of relevance since its focus lies in the translation of experience from external sources into epistemic cognition and meaning making on an individual basis. Kolb's theoretical perspectives are of importance when considering the processes of experiential learning, which in turn contribute to elements of subconscious bias that inevitably contribute to the professional identity of scientific educators from specific signature pedagogies and disciplines. Social learning theory also plays a significant role in the development of experiential learning. It is a perspective upon which health and medicine have drawn heavily upon in the shaping and framing of new pedagogical perspectives. It frames learning outcomes in the context of their impact in applied praxis rather than in terms of their individual impact and acknowledges that these are often a natural by-product of distinctive social cultures and contexts. This also aligns with the concept of lifelong learning within which there is the acknowledgement that scientific learning is dependent on levels of increasing experience (Hartman et al, 2020).

SCAFFOLDING LEARNING

XR provides an environment where the context of safety ensures that science learners can iteratively be scaffolded through processes of knowledge acquisition and the basis of psychomotor skill competency and eventual mastery can be gained and evidenced in scientific practice (Adefila et al, 2020).

One important aspect of the integration of XR though, is the degree to which it increases non-patient-based learning and the potential impact of this in relation to a lack of authentic human interaction across science-based health and medical curricula. What this debate highlights are the significance of strategic curriculum design and justification across all programmes, where pedagogical constructive alignment is pivotal to outcomes-based assessment techniques and the credibility of their implementation. Similarly, where variations and combinations of XR are integrated across academic and clinical curricula then the technology selected ought to be aligned with the need for the acquisition of specific skills and techniques. The integration of XR and its components does ensure, however, that during the process of learning and the acquisition of new knowledge, skills and behaviours that ensure:

- Aims and objectives can be aligned with the opportunity for ongoing feedback and the opportunity to rehearse and develop new scientific skill sets from novice to expert.
- Benchmarking of science learner performance can be isolated into specific learning domains and assessed accordingly, with the potential of identifying individual learning need.
- Interruption and hence unnecessary and unwanted distraction, during the process of science learning, can be avoided
- Within the context of health and medicine, patients can avoid any risk in terms of discomfort or the need to move regularly, as learners iteratively develop and refine new skills, where they can focus on domain specificity until a level of competency and eventual mastery is reached.
- That individual learning styles can be accommodated and encouraged by constructively aligning individual learning outcomes within each science teaching session, where time permits.
- That psychomotor tasks can be repeated so that exposure, practical viability and accuracy can be iteratively increased within the context of science.
- The opportunity for focused learner reflection on specific aspects of scientific knowledge and skill acquisition and a reflexive approach for the address of individually recognised shortfalls within the process of scientific learning.
- The theory-praxis gap can be gradually reduced by scaffolding science learning and gradually increasing exposure to real life scientific contexts where patients and their families are an integral part of authentic learning.

Non-technical skills such as interprofessional working and multidisciplinary team working and the clinical attributes they necessitate in terms of optimal interpersonal communication and the management of the clinical environment characterise the skill sets necessary to ensure the achievement of holistic learning in proficient scientific, healthcare and medical practice (Dehghani et al, 2021). Examples where this is now integrated in practice are in the skills and drills training, where it is also possible to integrate aspects of complex ambiguity and stress into situations, where a clear integration of psychomotor skill, affective and cognitive domain knowledge and application are necessary within any given scenario (DeMaria and Levine, 2013). It is these learning needs that high-fidelity clinical simulators address, so that once basic competence in a skill has been achieved then additional emotional stress and the need for instantaneous clinical decision making can be assessed. Whilst simulation is not an active replacement for the adrenaline rush faced, it does offer an insight within learner-centred experiential learning of the potentially uncertain scenarios that health and medical professionals often face in the real world. The portability of simulators such as 'SimMan' and 'SimMom' also provides context-specific training opportunities where situated learning can become an integral part of strategic curriculum justification, design and delivery (Viglialoro, 2021).

SUPPORTING CHANGE IN GLOBAL SCIENCE WORKFORCE PROVISION

Despite being one of the most needed and often underacknowledged parts of the global science workforce, most healthcare workers have been exposed to traditionalist mechanisms of education, which in terms of educational delivery, have remained pedagogically unaltered for the last half century (Karunathilake and Samarasekera, 2021). Extended reality as an integral part of a new technological age has been embedded within existing educational infrastructures, where traditional instruction mechanisms have shaped its implementation in relation to the pre-existing array of subject areas and academic disciplines represented by HEIs worldwide. Stemming from the traditionalist apprenticeship models of learning, these learning approaches have been shaken to the core by a global pandemic that despite being predicted, came as a complete shock to global academic communities where health and medical education had to continue, despite education as a social science being altered beyond recognition in a space of six months, when national lockdowns shaped altered approaches to educational affordance and opportunity to learn (Chan, 2021). Whilst medical publications have an average currency of five years, clinicians of health and medical practice work for an average of forty years on the foundational knowledge attained almost half a century before. It is perhaps no surprise to learn then, that medical errors have become a progressively iterative problem across global health and medical practice, in situations where technology has developed to such an extent that new interventions are often almost obsolete before their initial implementation at the front line of patient care (Melnyk et al, 2021). Whilst these issues are concerning, they also pose the greatest opportunity for processes of progressive change for health and medical students whose curricula will inform the complex and ambiguous territories of professionalism that they will need to negotiate over the forthcoming years. This is an opportunity not only to focus and stabilise the quality and optimal provision of education for the next generations of science and healthcare practice, but also to ensure that medical education becomes more accessible, cost effective and most importantly places patients at the heart of compassionate human care. Exponentially in response to these challenges there has been a progressive switch to Problem Based Learning pedagogical approaches to medical and healthcare curricula across the globe. The immediacy of feedback and impact of professional facilitation of science learning is central to the success of this approach in practice and ensures that the mindset of a critical enquirer is entrenched in learners who one day will become the next generation of scientific, medical and healthcare providers.

PATIENT AND CARER PUBLIC INVOLVEMENT IN SCIENTIFIC CURRICULUM DELIVERY

The integration of patients, their families and their carers within the context of patient and carer public involvement (PCPI) initiatives is now an integral part of most health and medical school curricula (Ocloo et al., 2021). Whilst they are invaluable in relation to the development of affective domain learning and interpersonal and effective communication skills, the fact that these skills are inseparable from highly attuned cognitive and psychomotor skills in practice still exists. Where they are therefore of particular value is in the integration of these skills in conjunction with high fidelity simulation equipment, so that psychological fidelity can also be integrated into the acquisition of psychomotor technical skills training (Johnson et al, 2020; Hayes and Graham, 2020). This is of value in the transitioning between specifically designed learning environments, within which simulated learning takes place, and the eventual 'real life' praxis that students face. In terms of recreating learning environments which also reflect the multidisciplinary and interprofessional working contexts within which health and medical staff find themselves daily, XR provides a useful mechanism of ensuring the psychological fidelity of scenarios used in learning. The origins of human factor analysis and historical emergence of XR training are well documented within the aviation industry, where contexts of flight simulation and the selective integration and conditions of flying can be reconstructed in terms of variability (Hawkins, 2017). One distinct variation in terms of the potential transferability of scientific learning to 'real life' though, is the complex ambiguity and unpredictability of working with humans, whose predictability far outweighs anything an aeroplane, however complex can reflect. Unlike flying, therefore, those assessments regarded as 'high stake' in terms of potential outcome and risk, cannot be conducted with medical simulation equipment because of their relative potential variability in pathological and physiological status (Spurgeon et al, 2019).

Using teaching and learning to drive assessment in science based academic disciplines is an acknowledged feature of student-centred learning but we must never lose sight of the fact that XR, whilst being an adjunct and embedded part of health and medical curricula must never become the driver of learning in itself, lest the most important learning opportunities and factors may be lost to the prioritisation of technology over human need. Similarly, some higher education institutions aspire to owning the latest equipment, which is often declared obsolete before it can be pedagogically integrated with any degree of credibility across those academic curricula for which it was originally purchased. The focus of curriculum design and how teaching and learning ought to drive assessment rather than vice-versa must be advocated if the overall resultant levels of quality in patient care are to remain optimal (Burgess et al, 2020).

EXTENDED REALITY AND ASSESSMENT MECHANISMS IN HEALTH EDUCATION

Within the context of the assessment of healthcare and medical students, formative assessment that is subject to the regulations of Professional and Statutory Regulatory Boards such as allied health and medicine programmes, there is also the opportunity to effectively moderate assessment within and between educational institutions in terms of the parity and equity of assessment mechanisms. Where clinical educators and academics have to evidence the objectivity of processes of assessment, then XR is invaluable in terms of being able to standardise approaches to both formative and summative assessment techniques and their respective outcomes in educational practice (Wilson and Shankar, 2021). The COVID-19 pandemic has seen an exponential rise in the delivery of blended synchronous online and face to face teaching, which has been termed a hybrid learning approach. However, a newer term is the Hyflex curricula which have permeated the educational marketplace, with tailored offers of individual flexibility in learning patterns, which can accommodate more diverse student cohorts. Instructional resources delivered by XR are often an integral part of these flexible offers and it can be an optimal way of managing learning time for those adult learners who have additional family or working commitments to fulfil (Gagnon et al, 2020). Operationally defined, Hyflex models also adopt a hybrid approach but offer the opportunity to study flexibly rather than at a pre-set time. Delineating between models of actual attendance on campus is often indicated by the term hybrid, whereas the actual curriculum model is termed blended.

CONCLUSION

Progressive innovation across scientific, medical and healthcare professions is something which challenges traditional, conservative approaches to pedagogy, which ultimately contributes to the life and death decision making of clinicians at the front line of care. In this regard, it is important that science and technology are seen as drivers of pedagogic practice, rather than methodologies in themselves. In this sense uptake of these new adjuncts is wholly dependent on medical and healthcare educators being open to change in approaches to education of subsequent generations of students and an acceptance that iterative and ongoing professional skills development is an integral part of adopting technology in practice. In practice this also necessitates the recruitment of staff with specialist technical skills who are responsible for the operationalisation of equipment, but who may not necessarily have specialist pedagogical skills in terms of curriculum design and implementation. So collaborative teamwork in the context of operational delivery of teaching sessions is pivotal. This investment, in terms of financial cost, collaborative working and mandatory iterative skills development for both academic and clinical educators, ensures that learning in a context safe from real-life risk with the potential for benchmarking the acquisition of knowledge and skills, is made possible in practice. The ongoing global COVID-19 pandemic has caused huge global disruption to traditional scientific, medical and healthcare education. Innovative responses in digital and technological tools have been quickly implemented to ensure that scientific learning can be maintained and sustained to ensure sufficient medical and healthcare graduates continue to qualify each year. Whilst change continues to ensure transformative approaches in those countries that can afford the fiscal implications of these developments, the gap between training providers globally also has the potential to widen in relation to optimal quality of medical and healthcare education. This is an obvious area to address if there is not to become a tangible difference in education for countries where the affordance of digital technology is not yet possible as standards for learners and their academic and clinical teachers. Recognising and addressing these key challenges can result in a truly authentic paradigm shift in the application of science based XR in pedagogic design, scholarship, and implementation.

REFERENCES

Abbas, J. R., Kenth, J. J., & Bruce, I. A. (2020). The role of virtual reality in the changing landscape of surgical training. *The Journal of Laryngology and Otology*, *134*(10), 863–866. doi:10.1017/S0022215120002078 PMID:33032666

Acharya, S., Bhatt, A. N., Chakrabarti, A., Delhi, V. S., Diehl, J. C., van Andel, E., & Subra, R. (2021). Problem-Based Learning (PBL) in Undergraduate Education: Design Thinking to Redesign Courses. In *Design for Tomorrow* (pp. 349–360). Springer. doi:10.1007/978-981-16-0119-4_28

Adefila, A., Opie, J., Ball, S., & Bluteau, P. (2020). Students' engagement and learning experiences using virtual patient simulation in a computer supported collaborative learning environment. *Innovations in Education and Teaching International*, *57*(1), 50–61.

Aebersold, M., & Dunbar, D. M. (2021). Virtual and Augmented Realities in Nursing Education: State of the Science. *Annual Review of Nursing Research*, *39*(1), 225–242. doi:10.1891/0739-6686.39.225 PMID:33431644

Agha, S. (2021). Aligning continuing professional development (CPD) with quality assurance (QA): A perspective of healthcare leadership. *Quality & Quantity*, 1–15.

Alnagrat, A. J. A., Ismail, R. C., & Idrus, S. Z. S. (2021, May). Extended Reality (XR) in Virtual Laboratories: A Review of Challenges and Future Training Directions. [). IOP Publishing.]. *Journal of Physics: Conference Series*, *1874*(1), 012031. doi:10.1088/1742-6596/1874/1/012031

Annala, J., Lindén, J., Mäkinen, M., & Henriksson, J. (2021). Understanding academic agency in curriculum change in higher education. *Teaching in Higher Education*, 1–18. doi:10.1080/13562517.2021.1881772

Antoniou, P., Arfaras, G., Pandria, N., Ntakakis, G., Bambatsikos, E., & Athanasiou, A. (2020). Real-time affective measurements in medical education, using virtual and mixed reality. In *International Conference on Brain Function Assessment in Learning* (pp. 87-95). Springer, Cham. 10.1007/978-3-030-60735-7_9

Argyris, C. (1991). Teaching smart people how to learn. *Harvard Business Review*, *69*(3).

Bevins, F., Bryant, J., Krishnan, C., & Law, J. (2020). *Coronavirus: How should US higher education plan for an uncertain future*. McKinsey.

Birt, J., Stromberga, Z., Cowling, M., & Moro, C. (2018). Mobile mixed reality for experiential learning and simulation in medical and health sciences education. *Information (Basel)*, *9*(2), 31. doi:10.3390/info9020031

Brandon, E., Freiwirth, R., & Hjersman, J. (2021, May). Special Session—Student Engagement with Reduced Bias in a Virtual Classroom Environment. In *2021 7th International Conference of the Immersive Learning Research Network (iLRN)* (pp. 1-3). IEEE.

Burgess, A., van Diggele, C., Roberts, C., & Mellis, C. (2020). Key tips for teaching in the clinical setting. *BMC Medical Education, 20*(2), 1–7. doi:10.118612909-020-02283-2 PMID:33272257

Chan, S. (2021). *Digitally Enabling'Learning by Doing'in Vocational Education: Enhancing' Learning as Becoming 'Processes*. Springer Nature. doi:10.1007/978-981-16-3405-5

Crouch, L., Rolleston, C., & Gustafsson, M. (2021). Eliminating global learning poverty: The importance of equalities and equity. *International Journal of Educational Development, 82*, 102250. doi:10.1016/j.ijedudev.2020.102250

Dehghani, M., Acikgoz, F., Mashatan, A., & Lee, S. H. (2021). A holistic analysis towards understanding consumer perceptions of virtual reality devices in the post-adoption phase. *Behaviour & Information Technology*, 1–19.

DeMaria, S., & Levine, A. I. (2013). The use of stress to enrich the simulated environment. In *The comprehensive textbook of healthcare simulation* (pp. 65–72). Springer. doi:10.1007/978-1-4614-5993-4_5

Dennick, R. (2016). Constructivism: Reflections on twenty-five years teaching the constructivist approach in medical education. *International Journal of Medical Education, 7*, 200–205. doi:10.5116/ijme.5763.de11 PMID:27344115

Durning, S. J., & Artino, A. R. (2011). Situativity theory: a perspective on how participants and the environment can interact: AMEE Guide no. 52. *Medical Teacher, 33*(3), 188–199. doi:10.3109/0142159X.2011.550965 PMID:21345059

Gagnon, K., Young, B., Bachman, T., Longbottom, T., Severin, R., & Walker, M. J. (2020). Doctor of physical therapy education in a hybrid learning environment: Reimagining the possibilities and navigating a "new normal". *Physical Therapy, 100*(8), 1268–1277. doi:10.1093/ptj/pzaa096 PMID:32424417

Gandolfi, E., Kosko, K. W., & Ferdig, R. E. (2021). Situating presence within extended reality for teacher training: Validation of the extended Reality Presence Scale (XRPS) in preservice teacher use of immersive 360 video. *British Journal of Educational Technology, 52*(2), 824–841. doi:10.1111/bjet.13058

Gerup, J., Soerensen, C. B., & Dieckmann, P. (2020). Augmented reality and mixed reality for healthcare education beyond surgery: An integrative review. *International Journal of Medical Education*, *11*, 1–18. doi:10.5116/ijme.5e01.eb1a PMID:31955150

Goh, P. S., & Sandars, J. (2020). A vision of the use of technology in medical education after the COVID-19 pandemic. *MedEdPublish*, 9.

Graf, A. C., Jacob, E., Twigg, D., & Nattabi, B. (2020). Contemporary nursing graduates' transition to practice: A critical review of transition models. *Journal of Clinical Nursing*, *29*(15-16), 3097–3107. doi:10.1111/jocn.15234 PMID:32129522

Hamilton, D., McKechnie, J., Edgerton, E., & Wilson, C. (2021). Immersive virtual reality as a pedagogical tool in education: A systematic literature review of quantitative learning outcomes and experimental design. *Journal of Computers in Education*, *8*(1), 1–32. doi:10.100740692-020-00169-2

Hartman, E., Reynolds, N. P., Ferrarini, C., Messmore, N., Evans, S., Al-Ebrahim, B., & Brown, J. M. (2020). Coloniality-decoloniality and critical global citizenship: Identity, belonging, and education abroad. *Frontiers: The Interdisciplinary Journal of Study Abroad*, *32*(1), 33–59. doi:10.36366/frontiers.v32i1.433

Hawking, S. (2009). *A brief history of time: from big bang to black holes*. Random House.

Hawkins, F. H. (2017). *Human factors in flight*. Routledge. doi:10.4324/9781351218580

Hayes, C., & Capper, S. (2020). Illustrating the transcendence of disciplinarity. In *Beyond Disciplinarity* (pp. 40–49). Routledge.

Hayes, C., & Graham, Y. (2020). *Designing a Benchmarking Tool for Testing Posttest Confidence Levels in Emergency Obstetrics Training*. SAGE Publications. doi:10.4135/9781529709285

Hayes, C., Hinshaw, K., & Petrie, K. (2019). Reconceptualising medical curriculum design in strategic clinical leadership training for the 21st century physician. In *Preparing Physicians to Lead in the 21st century* (pp. 147–163). IGI Global.

Heong, Y. M., Ping, K. H., Hamdan, N., Ching, K. B., Yunos, J. M., Mohamad, M. M., ... Azid, N. (2020). Integration of Learning Styles and Higher Order Thinking Skills among Technical Students. *Journal of Technical Education and Training*, *12*(3), 171–179.

Hilburg, R., Patel, N., Ambruso, S., Biewald, M. A., & Farouk, S. S. (2020). Medical education during the coronavirus disease-2019 pandemic: Learning from a distance. *Advances in Chronic Kidney Disease*, *27*(5), 412–417. doi:10.1053/j.ackd.2020.05.017 PMID:33308507

Hilty, D. M., Parish, M. B., Chan, S., Torous, J., Xiong, G., & Yellowlees, P. M. (2020). A comparison of in-person, synchronous and asynchronous telepsychiatry: Skills/competencies, teamwork, and administrative workflow. *Journal ofTechnology in Behavioral Science*, *5*(3), 273–288. doi:10.100741347-020-00137-8

Horton, S. (2021). Empathy Cannot Sustain Action in Technology Accessibility. *Frontiers of Computer Science*, *3*, 31.

Howell, H., & Mikeska, J. N. (2021). Approximations of practice as a framework for understanding authenticity in simulations of teaching. *Journal of Research on Technology in Education*, *53*(1), 8–20. doi:10.1080/15391523.2020.1809033

Humpherys, S. L., Bakir, N., & Babb, J. (2021). Experiential learning to foster tacit knowledge through a role play, business simulation. *Journal of Education for Business*, 1–7.

Jentsch, F., & Curtis, M. (2017). *Simulation in aviation training*. Routledge. doi:10.4324/9781315243092

Johnson, C. E., Kimble, L. P., Gunby, S. S., & Davis, A. H. (2020). Using deliberate practice and simulation for psychomotor skill competency acquisition and retention: A mixed-methods study. *Nurse Educator*, *45*(3), 150–154. doi:10.1097/NNE.0000000000000713 PMID:31246693

Jongerius, C., Hessels, R. S., Romijn, J. A., Smets, E. M., & Hillen, M. A. (2020). The measurement of eye contact in human interactions: A scoping review. *Journal of Nonverbal Behavior*, *44*(3), 1–27. doi:10.100710919-020-00333-3

Juraschek, M., Büth, L., Posselt, G., & Herrmann, C. (2018). Mixed reality in learning factories. *Procedia Manufacturing*, *23*, 153–158. doi:10.1016/j.promfg.2018.04.009

Kang, J., Diederich, M., Lindgren, R., & Junokas, M. (2021). Gesture patterns and learning in an embodied XR science simulation. *Journal of Educational Technology & Society*, *24*(2), 77–92.

Karunathilake, I. M., & Samarasekera, D. D. (2021). Learning In The 21st Century—'What's All the Fuss about Change?'. In Educate, Train and Transform: Toolkit on Medical and Health Professions Education (pp. 1-14). Routledge.

King, O., Borthwick, A., Nancarrow, S., & Grace, S. (2018). Sociology of the professions: What it means for podiatry. *Journal of Foot and Ankle Research, 11*(1), 1–8. doi:10.118613047-018-0275-0 PMID:29942353

Koufidis, C., Manninen, K., Nieminen, J., Wohlin, M., & Silén, C. (2021). Unravelling the polyphony in clinical reasoning research in medical education. *Journal of Evaluation in Clinical Practice, 27*(2), 438–450. doi:10.1111/jep.13432 PMID:32573080

Levitt, H. M. (2021). Qualitative generalisation, not to the population but to the phenomenon: Reconceptualising variation in qualitative research. *Qualitative Psychology, 8*(1), 95–110. doi:10.1037/qup0000184

Loeng, S. (2018). Various ways of understanding the concept of andragogy. *Cogent Education, 5*(1), 1496643. doi:10.1080/2331186X.2018.1496643

Logeswaran, A., Munsch, C., Chong, Y. J., Ralph, N., & McCrossnan, J. (2021). The role of extended reality technology in healthcare education: Towards a learner-centred approach. *Future Healthcare Journal, 8*(1), e79–e84. doi:10.7861/fhj.2020-0112 PMID:33791482

Luctkar-Flude, M., & Tyerman, J. (2021). The Rise of Virtual Simulation: Pandemic Response or Enduring Pedagogy? *Clinical Simulation in Nursing, 57*, 1–2. doi:10.1016/j.ecns.2021.06.008

Luo, C., Lan, Y., Luo, X. R., & Li, H. (2021). The effect of commitment on knowledge sharing: An empirical study of virtual communities. *Technological Forecasting and Social Change, 163*, 120438. doi:10.1016/j.techfore.2020.120438

Mathew, P. S., & Pillai, A. S. (2020). Role of Immersive (XR) Technologies in Improving Healthcare Competencies: A Review. *Virtual and Augmented Reality in Education, Art, and Museums*, 23-46.

McGrath, J. L., Taekman, J. M., Dev, P., Danforth, D. R., Mohan, D., Kman, N., ... Won, K. (2018). Using virtual reality simulation environments to assess competence for emergency medicine learners. *Academic Emergency Medicine, 25*(2), 186–195. doi:10.1111/acem.13308 PMID:28888070

Melnyk, B. M., Tan, A., Hsieh, A. P., Gawlik, K., Arslanian-Engoren, C., Braun, L. T., Dunbar, S., Dunbar-Jacob, J., Lewis, L. M., Millan, A., Orsolini, L., Robbins, L. B., Russell, C. L., Tucker, S., & Wilbur, J. (2021). Critical care nurses' physical and mental health, worksite wellness support, and medical errors. *American Journal of Critical Care, 30*(3), 176–184. doi:10.4037/ajcc2021301 PMID:34161980

Mitchell, R., & Boyle, B. (2021). Understanding the role of profession in multidisciplinary team innovation: Professional identity, minority dissent and team innovation. *British Journal of Management, 32*(2), 512–528. doi:10.1111/1467-8551.12419

Moreira, D. (2020). Virtual networks and asynchronous communities: methodological reflections on the digital. In *Ethnography in Higher Education* (pp. 177–196). Springer VS. doi:10.1007/978-3-658-30381-5_11

Mortimore, G., Reynolds, J., Forman, D., Brannigan, C., & Mitchell, K. (2021). From expert to advanced clinical practitioner and beyond. *British Journal of Nursing (Mark Allen Publishing), 30*(11), 656–659. doi:10.12968/bjon.2021.30.11.656 PMID:34109817

Negev, M., Dahdal, Y., Khreis, H., Hochman, A., Shaheen, M., Jaghbir, M. T., Alpert, P., Levine, H., & Davidovitch, N. (2021). Regional lessons from the COVID-19 outbreak in the Middle East: From infectious diseases to climate change adaptation. *The Science of the Total Environment, 768*, 144434. doi:10.1016/j.scitotenv.2020.144434 PMID:33444865

Obrad, C. (2020). Constraints and consequences of online teaching. *Sustainability (Basel), 12*(17), 6982. doi:10.3390u12176982

Ocloo, J., Garfield, S., Franklin, B. D., & Dawson, S. (2021). Exploring the theory, barriers and enablers for patient and public involvement across health, social care and patient safety: A systematic review of reviews. *Health Research Policy and Systems, 19*(1), 1–21. doi:10.118612961-020-00644-3 PMID:33472647

Okoye, K., Rodriguez-Tort, J. A., Escamilla, J., & Hosseini, S. (2021). Technology-mediated teaching and learning process: A conceptual study of educators' response amidst the Covid-19 pandemic. *Education and Information Technologies, 26*(6), 1–33. doi:10.100710639-021-10527-x PMID:34025205

Orr, N., Matthews, B., See, Z. S., Burrell, A., Day, J., & Seengal, D. (2021). Transdisciplinarity in extended reality (XR) research design: Technological transformation and social good (co-creation session at XR+ Creativity Symposium, University of Newcastle, 2020). *Virtual Creativity, 11*(1), 163-179.

Owens, K. P. (2021, July). Competency-Based Experiential-Expertise and Future Adaptive Learning Systems. In *International Conference on Human-Computer Interaction* (pp. 93-109). Springer, Cham. 10.1007/978-3-030-77873-6_7

Park, C., & Kim, D. G. (2020). Exploring the roles of social presence and gender difference in online learning. *Decision Sciences Journal of Innovative Education, 18*(2), 291–312. doi:10.1111/dsji.12207

Robert, I. V. (2021). Formation and development of digital transformation of domestic education on the basis of systemic convergence of pedagogical science and technology. In *SHS Web of Conferences* (*Vol. 101*, p. 03017). EDP Sciences. 10.1051hsconf/202110103017

Roussin, C. J., & Weinstock, P. (2017). SimZones: An organisational innovation for simulation programs and centers. *Academic Medicine, 92*(8), 1114–1120. doi:10.1097/ACM.0000000000001746 PMID:28562455

Silén, C., Wirell, S., Kvist, J., Nylander, E., & Smedby, Ö. (2008). Advanced 3D visualisation in student-centred medical education. *Medical Teacher, 30*(5), e115–e124. doi:10.1080/01421590801932228 PMID:18576181

Suryanti, S., Sutaji, D., Arifani, Y., Muyasaroh, M., & Zamzamy, M. (2020). Improved learning accessibility and professionalism of teachers in remote areas through mentoring development of teaching materials based on Augmented Reality. [Research Dissemination for Community Development]. *Kontribusia, 3*(1), 224–232. doi:10.30587/kontribusia.v3i1.1032

Tabatabai, S. (2020). COVID-19 impact and virtual medical education. *Journal of Advances in Medical Education & Professionalism, 8*(3), 140–143. PMID:32802908

Tang, K. S., Cheng, D. L., Mi, E., & Greenberg, P. B. (2020). Augmented reality in medical education: A systematic review. *Canadian Medical Education Journal, 11*(1), e81. PMID:32215146

Ten Cate, O., & Billett, S. (2014). Competency-based medical education: Origins, perspectives and potentialities. *Medical Education, 48*(3), 325–332. doi:10.1111/medu.12355 PMID:24528467

van der Niet, A. G., & Bleakley, A. (2021). Where medical education meets artificial intelligence: 'Does technology care?'. *Medical Education, 55*(1), 30–36. doi:10.1111/medu.14131 PMID:32078175

Viglialoro, R. M., Condino, S., Turini, G., Carbone, M., Ferrari, V., & Gesi, M. (2021). Augmented Reality, Mixed Reality, and Hybrid Approach in Healthcare Simulation: A Systematic Review. *Applied Sciences (Basel, Switzerland), 11*(5), 2338. doi:10.3390/app11052338

Voštinár, P., Horváthová, D., Mitter, M., & Bako, M. (2021). The look at the various uses of VR. *Open Computer Science*, *11*(1), 241–250. doi:10.1515/comp-2020-0123

Vygotsky, L. S. (1978). Zone of proximal development: A new approach. *Mind in society: The development of higher psychological processes*, 84-91.

Welie, J. V. (2004). Is dentistry a profession? Part 3. Future challenges. *Journal - Canadian Dental Association*, *70*(10), 675–678. PMID:15530264

Wilson, I., & Shankar, P. R. (2021). The COVID-19 pandemic and undergraduate medical student teaching/learning and assessment. *MedEdPublish*, *10*(1), 10. doi:10.15694/mep.2021.000044.1

World Health Organisation. (2021) International Data Online Updates. WHO. https://www.who.int/data)

Yigitcanlar, T., Butler, L., Windle, E., Desouza, K. C., Mehmood, R., & Corchado, J. M. (2020). Can building "artificially intelligent cities" safeguard humanity from natural disasters, pandemics, and other catastrophes? An urban scholar's perspective. *Sensors (Basel)*, *20*(10), 2988. doi:10.339020102988 PMID:32466175

Zulkifli, A. F. (2019). Student-centered approach and alternative assessments to improve students' learning domains during health education sessions. *Biomedical Human Kinetics*, *11*(1), 80–86. doi:10.2478/bhk-2019-0010

ADDITIONAL READING

Cranmer, K., Brehmer, J., & Louppe, G. (2020). The frontier of simulation-based inference. *Proceedings of the National Academy of Sciences of the United States of America*, *117*(48), 30055–30062. doi:10.1073/pnas.1912789117 PMID:32471948

Hong, T., Langevin, J., & Sun, K. (2018, October). Building simulation: Ten challenges. [). Springer Berlin Heidelberg.]. *Building Simulation*, *11*(5), 871–898. doi:10.100712273-018-0444-x

Jeffries, P. (2020). *Simulation in nursing education: From conceptualisation to evaluation*. Lippincott Williams & Wilkins.

Kadian, A., Truong, J., Gokaslan, A., Clegg, A., Wijmans, E., Lee, S., Savva, M., Chernova, S., & Batra, D. (2020). Sim2Real predictivity: Does evaluation in simulation predict real-world performance? *IEEE Robotics and Automation Letters*, *5*(4), 6670–6677. doi:10.1109/LRA.2020.3013848

Satin, A. J. (2018). Simulation in obstetrics. *Obstetrics and Gynecology*, *132*(1), 199–209. doi:10.1097/AOG.0000000000002682 PMID:29889745

KEY TERMS AND DEFINITIONS

Augmented Reality (AR): Is a technology capable of superimposing or overlaying a computer-generated image across a visual projection of the real world, providing a composite view of the two.

Extended Reality (XR): Is the term given to all real-and-virtual combined environments and human-machine interactions, which are essentially digitally generated by technology and wearable accessories.

Hyflex Curriculum: Is an adaptation of hybrid learning where each class session and learning activity is offered in-person, synchronously online, and asynchronously online. With this approach learners make the decision of how they will participate with the learning opportunities afforded to them.

Immersion Technology: Is the digital equipment which provides the perception of being present in a created and non-physical world.

Mixed Reality (MR): Is the merging of virtual and actual reality to provide new mechanisms for visualising given scenarios. The physical objects and digital objects can interact with each other in real time.

Simulation: Is the integrated use of a computer model, which imitates reality in the context of study, where risk can be eliminated as part of initial scaffolded learning.

Virtual Reality (VR): Is the digitally generated simulation of a 3-D image or situated context or learning environment, within which a learner can be placed and with which they can interact by wearing electronic accessories such as eye goggles or gloves with sensors.

Chapter 3
Competency-Based Assessments in STEM Classrooms:
Using Creative Story Writing as a Formative Assessment

Garima Bansal
https://orcid.org/0000-0002-8853-1708
Australian Council for Educational Research, India

ABSTRACT

It has often been observed that what is assessed often becomes what is taught. To enable teachers to teach 21st century skills, it is crucial that school systems upskill their models of designing high-quality competency-based assessments. This chapter reports a competency-based formative assessment used to equip students with the skills required for successful survival in the 21st century. It reports a formative assessment task in which story writing is used to develop graph development skills among STEM students studying in middle grades in an Indian school classroom. The results indicate that competency-based assessments unravel student-held misconceptions which can be used by the teachers to provide appropriate formative feedback.

DOI: 10.4018/978-1-6684-6932-3.ch003

The principal goal of education is to create people who are capable of doing things, not simply of repeating what other generations have done - people who are creative, inventive, and discoverer. (Jean Piaget as cited in Ginsberg & Opper, 1969, p. 5)

THE LEARNING IMPERATIVES OF NEW SOCIETY: NEED FOR COMPETENCY-BASED ASSESSMENTS

Recent years have witnessed a change in the socio-political-economic landscape of the world with the surfacing of neo-liberalisation, globalisation, migration of people on varying accounts, ongoing and resurgent violent conflicts, and outbreak of health epidemics like COVID-19 to list a few. These changes are bound to impact school life which serves as a springboard for crystallising social values. More than ever, now students are finding it important to focus on competencies including critical thinking, communication, collaboration, and creativity in order to successfully navigate through the demands and challenges of the local as well as the global environment. National education policies, school curricula and large-scale assessments across the globe are swiftly shifting gears towards upskilled teaching and assessing competencies (Care, Griffin, & Wilson, 2018).

With the dawn of 'knowledge era' and an advent of 'learning age', educational objectives are rapidly changing, and educational institutions are bestowed with the responsibility of inculcating among its pupils a wider range of dispositions than just academic knowledge and skills. According to data published by Global Business Coalition for Education, the Education

Commission and UNICEF, over half of Indian students will leave school without the necessary skills to get 21st century jobs by 2030. This is so as employers are now looking for employees not with high qualifications but people who possess problem solving skills even if they are high school dropouts (NCERT, 2006, p.5). This scenario is common across the world as indicated by a Gallup study of Americans aged 18–35. It emerged that "those with high 21st century skill development is twice as likely to have higher work quality compared to those who had low 21st century skill development" (Gallup 2013, p. 4).

Global employment market seeks skills like meta-cognition- ability to think about one's own thinking. It includes a variety of self- awareness processes to help plan, monitor, orchestrate, and control one's own learning. Creativity is another desired skill for the present day technologically advanced world. It is essential to solve newer problems and generate ingenious ideas that can serve the expanding and more technically sophisticated needs of industry and business (McKinsey Global Institute, 2017).

For this, our school system needs to provide students with educational experiences that support the establishment of linkages of school life with the larger world outside, and promote divergent ways of thinking and problem solving, in short, 'knowing what to do when you don't know what to do' (Claxton, 2002 as cited by Broadfoot, 2008, p.46).

In this chapter, I explore the ways in which competency-based assessments can be designed to support development of scientific literacy in the context of an Indian classroom.

TIME FOR CHANGE: A MOVEMENT TOWARDS COMPETENCY-BASED ASSESSMENTS

Various educational policy documents have been reiterating that in the present day society vast amount of information is easily accessible at the click of a mouse through the use of internet, so the need of the hour is no more gathering factual information rather it is the ability to sift, select, and organize the information and knowledge meaningfully to solve problems and make decisions in new situations (NCERT, 2006). In his prologue to the book "Four Dimensional Education" Andreas Schleicher (p.1) says,

Put simply, the world no longer rewards people just for what they know—search engines know everything—but for what they can do with what they know, how they behave in the world, and how they adapt. Because that is the main differentiator today, education is becoming more about creativity, critical thinking, communication, and collaboration; about modern knowledge, including the capacity to recognize and exploit the potential of new technologies; and, last but not least, about the character qualities that help fulfilled people live and work together and build a sustainable humanity (Fadel, Bialik, & Trilling, 2015).

Concerted efforts have been laid by global West over last decade to ensure that teaching-learning and assessment practices support the development of twenty-first century skills (21C, henceforth) (Schleicher, 2012). To name a few, The Partnership for Twenty-First Century Skills (P21) provided directions for high-quality skill-based teaching in sixteen US states (P21, 2009); and the international ATC21S—Assessment and Teaching of Twenty-First Century Skills guided changes in assessment practices in line with 21C skills.

Now, the question emerges that why despite the global recognition of the importance of 21C skills as an essential element of schooling students are still not learning it (Saavedra & Opfer, 2012). To which, they offered an explanation that teachers are not teaching them. It is a fact that teaching-learning practices are still rooted in transmission model which does not support the development of these skills. Also, teachers lack specific expertise and deep training required for teaching these skills in their classrooms (Comfort & Timms, 2018). Furthermore, traditional assessments pose hindrance to teachers giving enough time to teach these skills to their students. They consider it a waste of time as assessments do not seek to evaluate a student's understanding on these skills (Broadfoot, 2008). Therefore, the need of the hour is to set up an educational model that serves the demand of the changing world.

Research notes that any education system is designed to serve then existing social and economic needs of the times. And, assessment necessarily functions as a mediating link between school and society (Bansal, 2014). While tracing the origins of formal examination, it was found that testing served the societal needs of selection and certification in order to ration educational opportunity for deserving few. While in the present day, social and economic demands are significantly different from the past. There is a movement from a world in which educational opportunity had to be rationed to a world in which educational opportunities are plentiful (Broadfoot, 2008). Therefore, it comes to fore that urgent changes are necessitated in the educational processes and priorities.

With changes in educational goals, there is a parallel need of developing assessment apparatuses that are supportive and encouraging, flexible and transparent. Renowned education assessment researcher Patricia Broadfoot observed that employers are looking for individuals who can think, solve problems and work in teams, but we are increasingly subjecting our children to tests that require cramming, reproduction of existing knowledge and individual competition (Broadfoot, 2008). One of the major accusations levied on formal testing is that it promotes rote memorisation of factual content depreciating the importance of understanding the concept which involves the ability to organise knowledge, to relate it actively to new and past experiences, and form 'big ideas' which can distinguish a novice from expert (Harlen, 2008). Big ideas are those that can be applied in different contexts; enable learners to understand a wide range of phenomena by identifying the essential links between different situations without being diverted by superficial features.

Broadfoot (2008) further argued for the dire need of recognizing the incompatibility between traditional high- stakes tests and twenty first century educational demands. She reiterated that in today's world the aim of educational system is not only selecting the deserving few to reach the pinnacle of educational opportunities rather the challenge lies in creating lifelong learners who are motivated enough to develop themselves throughout their lives. She suggested that assessment that supports learners both in formal education and beyond, must be 'sustainable' in a way it encourages learning throughout life in formal and in informal settings. It has been suggested in policy documents that using outmoded assessment technologies to assess significantly different educational priorities eventually results in creation of 'unemployable' youth (World Economic Forum, 2018).

Harlen (2008) while citing OECD (1999) pointed out that, 'students cannot learn in school everything they will need to know in adult life. What they must acquire is the prerequisites for successful learning in future life' (p.9). Effective learning experiences along with helping students to think and reflect on their learning practices seem integral to authentic learning and this is where a framework of educational assessment which is responsive to learners' learning needs is located. Assessments situated in real-world, authentic environments that enable students to build on existing mental models and apply their knowledge to solve real-world problems is the need of the day (Trilling & Fadel, 2012).

COMPETENCY-BASED ASSESSMENTS

Assessment often enjoys the hegemonic control over all other education processes (Broadfoot, 2008). Therefore, for development of 21C skills among school children, it is imperative to introduce a timely change in the assessment processes. One of the ways is to use international large-scale assessments, such as, PISA, TIMSS etc. that use competency-based questions to gauge health of a country's education system. Results from these assessments can be reliably used to inform education policy and lead to consequent education reform (Care, 2020). The other approach is to introduce competency-based formative assessments that could meaningfully inform classroom-based teaching-learning practices (Black & Wiliam, 1998).

Competencies are defined as "a combination of skills, abilities, and knowledge needed to perform a specific task" (Jones, Voorhees, & Paulson, 2002, p.viii). According to the OECD (2005), a competency is

more than just knowledge and skills. It involves the ability to meet complex demands, by drawing on and mobilising psychosocial resources (including skills and attitudes) in a particular context. For example, the ability to communicate effectively is a competency that may draw on an individual's knowledge of language, practical IT skills, and attitudes towards those with whom he or she is communicating (p.4).

A single competency can be used in different ways by different professionals. To illustrate, both professional golfers and surveyors skilfully make use of the competency of measuring distances. So, to say that competencies developed in one domain may find its' application and relevance in other domains. Thus, competencies are cross-disciplinary in nature. Similarly, one competency may involve different set of sub-skills when applied in different contexts. To illustrate, leadership in surgery is different from leadership on the basketball court. Superior knowledge of the surgical procedures is more important to a surgeon while ability to motivate teammates is crucial to a leader in a basketball court. However, few skills form the basics of leadership in both the diverse contexts: ability to effectively coordinate the roles, timing, and contributions of co-workers etc. (Jones, Voorhees, & Paulson, 2002). The bundles of different skills and knowledge that are given the same label in different contexts necessitates the development of a common understanding of competencies.

Many frameworks have been globally developed to define 21C competencies: "Framework of Future Competences" developed by UNESCO's International Bureau of Education" (UNESCO, 2016), OECD learning compass, a product of the Organisation of Economic Cooperation and Development "Future of Education and Skills, 2030" project (OECD,2017), ACER general capabilities framework (Scoular, Ramalingam, Duckworth, & Heard, 2020). Similarly, scholars like Charles Fadel, Maya Bialik and Bernie Thrilling provided a Four Dimensional Education Framework in the book Four Dimensional Education (Fadel, Bialik, & Trilling, 2015). This framework serves as a guiding framework that can help to re-examine school curricula for 21st century education. It contains four dimensions: knowledge (what we know and understand), skills (how we use knowledge – what we do with what we know), character (how we behave and engage in the world), and meta-learning (how we reflect and adapt).

Competency-based assessments encourage learners to apply their understandings to different authentic contexts. This approach enables them to translate theoretical underpinnings developed in classrooms to problem-solving situations found in the real-world. These assessments provide a mental framework to children which they can adopt when faced with similar problems in their daily lives. Also, another important aspect of CBA is its' ability to uncover student-adopted algorithms which they use to arrive at solutions to the problems posed in CBAs. It is crucial for teachers to develop insights about these algorithms so that teachers can provide child-centric instructional correctives to individual learners (Bansal, 2020).

In line with various competencies described in the above-mentioned frameworks, Harlen (2008) pointed out that the conception of 'literacy' has undergone a major change in the twenty first century. It is no more understood as uni-dimensional ability to read and write but interpreted as a cluster of multi-dimensional abilities such as 'technological literacy', 'scientific literacy', 'mathematical literacy', 'political and social literacy' so on and so forth. Since scientific literacy encourages individuals to know about scientific concepts and processes and ways of using this information for a sustainable living in the environment, it is regarded as a key competency (Turiman, Omar, Daud, & Osman, 2012).

SCIENTIFIC LITERACY

Scientific literacy was first described by Paul DeHart Hurd in 1958. Since then different researchers and reformists have defined scientific literacy in various ways responding to the demands of the changing social-political world. George DeBoer (2000) presented a historical review of the term scientific literacy. Broadly speaking, he noted that all the definitions of scientific literacy encapsulate the goal of science education as nurturing scientifically literate students who are not just aware of the conceptual knowledge of scientific theories and principles but also possess a sound understanding of scientific methods. National Research Council defines scientific literacy as "the knowledge and understanding of scientific concepts and processes required for personal decision making, participation in civic and cultural affairs, and economic productivity" (NRC, 2012). It is imperative that an educated child or adult develops scientific literacy to distinguish facts from fiction, to gauge the authenticity of scientific methods used to arrive at solutions to the problems, to analyze varying competitive explanations provided about societal and scientific events, and propose ways to scientifically investigate natural phenomena (OECD, 2019).

The NRC framework broadens the definition of scientific literacy from the ability of simply knowing scientific concepts to the ability of a student to engage in the following scientific practices: asking questions and defining problems; developing and using models; planning and carrying out investigations; analysing and interpreting data; using math, information and computer technology, and computational thinking; constructing explanations and designing solutions ; engaging in argument from evidence; and obtaining, evaluating and communicating information (NGSS Lead States 2013). Pellegrino observes a strong overlap between science standards and 21C skills (NRC, 2012). National Research Council framework emphasises the development of critical thinking, collaborative problem solving, constructing and evaluating evidence-based arguments, systems thinking, and complex communication as crucial to science and engineering related disciplines (NRC, 2012). On similar lines, TIMSS assessment framework highlighted science practices such as asking questions based on evidence, generating evidence, working with data, answering research questions and making arguments from evidence are practices that need to be assessed in the context of science domain (TIMSS Assessment Framework, 2019). These competencies cannot be taught or tested in isolation and need to be embedded in the pedagogical and assessment approaches (CBSE, 2021).

Researchers have argued that it is imperative to purposefully teach scientific literacy right from early years to higher grades. It's been argued that scientific literacy has different degrees and forms. It expands and deepens over a lifetime. However, it is crucial to lay a strong foundation of scientific literacy during early years as the attitudes and values established toward science in the early years are going to have a strong influencing factor on a person's development of scientific literacy as an adult. At early years, a basic understanding of the natural phenomena is essential to survive in a healthy manner in the society. For example, even young learners should possess the ability to evaluate different explanations provided by the adults regarding different natural phenomena, such as, "chickens are not birds" or "don't hold a worm because it will bite you" etc. (Ashbrook, 2020).

Importance of developing scientific literacy is furthered at adulthood when they encounter various societal issues, such as, COVID-19 pandemic, or other routine issues of making sense of scientific information featuring in newspapers, magazines etc. (Zucker, 2021). He emphasised five themes for developing students' scientific literacy: personal and societal contexts of science, relating scientific literacy with other forms of literacy, distinguishing reliable scientific information from junk information, explicit teaching of nature of science, and providing inclusive environment for females and minorities to grow in science.

Display of scientific literacy varies in different ways in different individuals, such as, appropriate use of scientific terminologies to describe scientific phenomena behind various natural and societal events or applying scientific concepts and processes to arrive at solutions to problems encountered in daily lives. In addition, individuals may demonstrate differences in literacy in different domains, such as some may possess a nuanced and detailed understanding of life-science concepts and words, and less understanding of physical-science concepts and words or vice-versa.

Program of International Student Assessment (PISA) has identified three critical science-specific competencies (see Table 1).

Table 1. PISA scientific literacy

Competency	Skills
Explaining phenomena scientifically	Recognising, offering and evaluating explanations for a range of natural and technological phenomena through: Recalling and applying appropriate scientific knowledge; Identifying, using and generating explanatory models and representations; Making and justifying appropriate predictions; Offering explanatory hypotheses; Explaining the potential implications of scientific knowledge for society
Evaluating and designing scientific enquiry	Describing and appraising scientific investigations and proposing ways of addressing questions scientifically through: Identifying the question explored in a given scientific study; Distinguishing questions that are possible to investigate scientifically; Proposing a way of exploring a given question scientifically; Evaluating ways of exploring a given question scientifically; Describing and evaluating a range of ways that scientists use to ensure the reliability of data and the objectivity and generalisability of explanations.
Interpreting data and evidence scientifically	Analysing and evaluating scientific data, claims and arguments in a variety of representations and drawing appropriate conclusions through: Transforming data from one representation to another; Analysing and interpreting data and drawing appropriate conclusions; Identifying the assumptions, evidence and reasoning in science-related texts; Distinguishing between arguments that are based on scientific evidence and theory and those based on other considerations; Evaluating scientific arguments and evidence from different sources (e.g. newspaper, internet, journals).

(Source: OECD, 2019, pp.104-5)

The case study described in this chapter presents the use of stories as a competency-based formative assessment task aimed at developing students' scientific literacy. Studies using stories as an assessment tool have been reviewed in the next section.

Stories in STEM

Schiro and Lawson (2004) observed that stories transform the abstract, objective, and deductive mathematical world into a subject surrounded by imagination, subjective meaning, and feelings. Since then, storytelling technique has gained prominence as a pedagogical approach in STEM education. Casey and colleagues (2008) used stories for fostering preschool children' geometry skills. Likewise, Goral and Gnadinger (2006) used a fantasy story to expand grade one students' familiarity with the concept of place value in the USA. In Greece, Lemonidis and Kaiafa (2019) used this technique to teach fractions to the third-grade students. Balakrishnan (2008) presented nine unique stories for teaching abstract mathematical concepts, such as, irrational numbers, Pythagoras theorem, exponents etc. to secondary grade learners. All these studies reinforce the importance of storytelling as a stimulating ground for developing ever-lasting learning of difficult concepts by students. This is so as students often get emotionally absorbed in solving problems of the characters in the story. This heightened motivation lead to students' engagement in dialogic conversations that require them to work towards the big ideas and generalise the concept beyond the immediate story context. Albano and Pierri (2014) used Raymond and Ohler's Storytelling Design Model in their research project "Obiettivo 500". Protagonist of their mathematical story was a journalist who was expected to predict election result using different forms of graphs. This story encouraged graph interpretation as the fundamental mathematical capability to be developed among learners. Lacefield and Markert (2019) used creative writing to enhance students' understanding of mathematical concepts. They observed that stories provide an authentic context to students wherein they get freedom to reformulate and review their mathematical thinking.

Similarly, there is a long tradition of using stories in science classrooms for distinct purposes: making science more interesting and engaging, illustrating a concept, providing opportunities for contextualized learning, making content memorable, integrating science and literature, use of biographic accounts of scientists to foster empathy in students about the struggles, challenges and experiences of scientists, introduce nature of science, foster inquiring mindset among learners, to introduce thought experiments, and raise awareness about environmental issues (Hadzigeorgiou, 2016). It is an important tool to arise curiosity in learners, support them in problem posing and problem solving, and think of creative solutions to problems (Klassen & Froese-Klassen, 2014). The present case study uses the power of story-writing in a cross-cutting domain of STEM.

THE CASE STUDY

This article reports the use of story-writing as a competency-based assessment in middle grade classroom in an urban school located in one of the cosmopolitan cities in India. This is an action research conducted by the author herself.

Presenting Classroom Scenario

While I was teaching graphs in STEM classrooms, it emerged that students were finding it boring and facing difficulties in applying the graph understanding to their real-life situations. This observation is backed by research studies conducted in different parts of the world that clearly suggest that contextualizing graph instruction in classrooms is crucial in ensuring that learners' prior knowledge structures and newly acquired knowledge of graph interpretation and construction is translated in solving real world problems (Glazer 2011; Friel, Curcio, & Bright, 2001). To understand the impact of context-familiarity on students' ability of graph interpretation, Shah and Carpenter (1995) conducted an experimental research. They compared viewers' graph interpretation skills on two forms of data sets: familiar data (such as number of car accidents, number of drunk drivers, and traffic density), and unfamiliar data sets (e.g. ice-cream sales, fat content, sugar content). It emerged that in familiar contexts, viewers tended to describe general trends ignoring the idiosyncratic data points inconsistent with general trends, while in unfamiliar contexts their observation was restricted to identification of local maxima and minima. Thus, highlighting the fact that context-familiarity is supportive to students' ability to effectively draw patterns from graphs. According to Wallace et al. (2004), one way to gauge scientific practice, critical thinking and collaborative problem solving is through writing. When students engage in the process of science writing, their ideas progress from vague notions to more complex understandings (Keys et al. 1999); and writing to enhance learning becomes more powerful when combined with collaborative peer discussion (Chen et al. 2013).

This study extends the research from graph interpretation to graph construction using *story writing* as a competency-based formative assessment task. *Formative Assessment* is an integral part of teaching-learning (Black & Wiliam, 1998). Wiliam and Thompson (2007) conceptualized the following as the key strategies of formative assessment:

- clarifying the learning intentions and criteria of success
- engineering effective classroom discussions and other learning tasks
- providing constructive feedback,
- activating students as instructional resources for one another
- activating students as the owners of their own learning

To this end, Heritage (2010) adds the importance of co-construction of learning criteria of the assessment task with students. She suggests that co-construction of assessment criteria enables learners to internalize the notion of 'quality' work. It supports them in reviewing their own work with reference to the assessment criteria of the task, thus, encouraging the practice of self- assessment. Research of formative feedback reiterates that it should be task-specific giving clear instructions to students about ways to improve their performances rather than making comparisons with other students (Hattie & Timperley, 2007).

After teaching the preliminaries of graph construction on Day 1 and 2, such as, skills to choose appropriate scale, plot coordinates, provide index, draw axes, and providing nomenclature to the plot curves, I engaged students in a discussion around estimation of distance and time on Day 3. Questions such as, can you guess the approximate distance between your table and teacher's table, distance between your city and Shimla and so on were initiated. Similarly, questions were initiated to encourage students to estimate the time taken to reach from home to school, time taken to reach Chennai by car or by flight etc. to foster students' estimation skills.

To foster graph comprehension, I asked students to collect different graphs from newspapers, magazines, textbooks etc. as a home task on Day 3. On Day 4, the classroom community created a common pool of all the graphs brought by students. Students were divided into different groups having 3-4 students per group. And, each group was encouraged to pick up a graph from the pool. They were given enough time to interpret the graph and present their understanding to the entire class. This activity encouraged students to draw inferences from different types of graphs available all around them. Thereby, increasing their ability to apply their conceptual understanding to real-life situations.

Researchers have observed that along with developing graph comprehension among students, it is imperative to make them 'inventers of displays that convey their own messages about the meanings of the data they are using' (Friel, Curcio and Bright, 2001, p. 150). Hence, the obvious next step that followed graph comprehension exercise was to engage students in tabulation of data and representation of data in graphical manner on Day 5 by the way of a formative assessment task.

The Formative Assessment Task

The formative assessment task was designed to encourage students to create graphs to represent their real-life situations. Being a competency-based assessment task, the intent was to enable students to understand the importance and usability of graphical representations in authentic life contexts.

The task involved writing a story involving motion of two characters, tabulating the mathematical data involved in the story, and plotting the data points in a distance–time graph. Assessment criteria were developed with the students in the classroom itself that were mapped on to the PISA scientific assessment competency framework to arrive at the following:

Evaluating and designing scientific enquiry

- Write a story involving motions of two characters

Interpreting data and evidence scientifically

- Communication of the mathematical data in a tabular format

- Plotting of distance-time graph demonstrating motion of characters
- Graph drawn should have an appropriate scale, co-ordinates should be marked accurately, index to distinguish between the motions of two different characters

Explaining phenomena scientifically

- Ensuring congruence among verbal and visual descriptions

Student Misconceptions

Student work elicited different misconceptions i.e. inappropriate or faulty understanding of underlying scientific and mathematical concepts and processes which are often elicited when students are confronted with problem-solving situations. This section presents dominant student-held misconceptions with respect to PISA competencies along with student work samples.

Evaluating and designing scientific enquiry

The competency of *explaining phenomena scientifically* necessitates students to recall the appropriate content knowledge applicable in a given situation and use it to generate tentative explanatory hypotheses for an observed phenomenon. People possessing this competency should be able to make meaning out of the data using standard scientific models, construct simple representations for everyday phenomena and use these representations to make predictions. This competency further involves the ability to describe or interpret phenomena and identify appropriate descriptions, explanations, and predictions (PISA, 2018).

This task encouraged students to write a story describing motion of two characters. Students' data demonstrated several misconceptions. For example, some of the student stories failed to describe motion of characters in terms of mathematical data, or partial data was presented either for distance or for time, or unrealistic data for distance and time was given. Illustrations of student work samples are provided in Table 2.

Table 2. Misconceptions in student stories

Misconception: No mathematical data in the story
Student work sample Concho Frog called all the animals of the forest to a meeting. Once they all gathered, he made an announcement: "There is going to be a skateboard race next Sunday. It will be at ten o' clock in the morning. All of you are welcome to register for the race. There will be a prize for the winner…" "What will the prize be?" asked a bird called Crane. "The prize will be a trip around india with all the expenses paid answered Mr. Concho Frog. "I want to get the prize," said the red tail hawk. "We shall see who wins the race," replied the sparrow. "I have already started training for the match," said the owl to the lizard. They all talked to one another, each one claiming to be the winner of the race. The meeting lasted until it was time to go home. The next morning, at sunrise, the forest was filled with voices and skateboard sounds. Some were riding their skateboards, while others were oiling the wheels and repairing the board. They all were very busy and excited. Well, most of them, because the parrot neither has a skateboard nor did she know how to ride. The parrot heart was full of sorrow. She was alone on her nest when she heard the call of the hummingbird. "Hi, parrot," said the hummingbird. "Hi, how are you? She answered in a soft, sad little voice. What's wrong? why aren't you practicing for the race? The parrot told her grief to the hummingbird. She was so sad that her eyes were filled with tears. "Come on, parrot, don't cry. We have plenty of time to prepare for the race. We are six days ahead. I will lend you my skateboard and teach you how to use it." said the hummingbird. "But, what about you"? asked the parrot. "I can't be on the race that day because I have to be with my wife. She is hatching three eggs and, you know…" replied The hummingbird. "You are really a good friend!" said the parrot with joy. With love, patience and care the hummingbird trained the parrot to ride the skateboard. The day before the race he told her: "I see that you have learned to ride the skateboard quite well. Now, the big secret! I will show you how to ride it faster." "Good. because I really need it." "Look, all you need to know is how to move your wings like this." We all know that the hummingbirds are expert moving their wings. Thus, we can imagine that the hummingbird taught her well. The day of the race there were four competitors. the biggest was the red tail hawk and the smallest was the ladybug. They were ready and set. The go signal was given. A cheer went up and They all started running as fast as they could. The ladybug took the lead. At the end of the first curve, the parrot, which was running fourth, saw how the crane bumped the ladybug with his wing. She was pushed away from first to third place. Then the owl shoved her away from the track on purpose. The parrot stopped to see if the ladybug was hurt. "It's not fair! Races should be won fairly", said the parrot while helping the ladybug. "Are you OK? " questioned the parrot. "I am fine, It's just that I am too small…" said the ladybug almost crying. "Cheer up and let's go back to the track. The race has not finished yet! I will help you", said the parrot. They got back on the track. The parrot helped the ladybug gain speed by flapping her wing behind her. One after the other the contestant were left behind. Right at the finish line a sudden zoom wiped the crane's smile from his face. It was the ladybug, passing by his side at speed of light. Soon, Mr. Concho Frog announced the order in which the runners finished. "And the winner is... the ladybug! Second place goes to the parrot, Third place to the crane, Fourth place…" Well, let's just skip the arriving order and see how the story ends. "Thank you, parrot, for helping me win the skateboard race. Without your help I would not have made it", said the ladybug. "It was a pleasure", replied the parrot, with a smile, while she remembered how somebody had helped her as well. "Come on, I want you to visit a friend of mine." They went to the hummingbird's nest. There they found him and his wife pleased with the arrival of their birdies.

Interpreting data and evidence scientifically

The competency of explaining phenomena scientifically entails skills required to evaluate scientific reports and investigations critically, identify scientifically investigable questions, propose ways of exploring investigable questions in scientifically valid ways, describe and evaluate a range of ways that scientists use to ensure the reliability of data and the objectivity and generalisability of explanations. This competency requires fundamental knowledge of the key features of a scientific investigation that includes ways of identifying what things should be measured, what variables should be changed or controlled, and what action should be taken so that accurate and precise data can be collected (PISA, 2018).

This task encouraged the learners to transform data from one form to another. Specifically, it encouraged learners to communicate the mathematical data presented in the story in a tabular format, plot distance-time graph demonstrating motion of characters presented in the story, and the graph drawn should have an appropriate scale, co-ordinates should be marked accurately, index to distinguish between the motions of two different characters.

Tabular representation of data presented in stories involved various errors: physical quantities of time and distance were presented as cardinal quantities or discrete variables whereas for plotting of distance-time graphs it should be presented as continuous variables (Zevenbergen, Shelley, & Wright, 2004). Other errors were inconsistencies between units of distance and time mentioned in the story and in the table. Some student stories did not include any data but it was included in the table. A few students did not mention either the titles or units of the physical quantities being represented by the columns in the table which made it incomprehensible for graph plotting purposes. In other cases, names of the characters whose data was being included in the table was missing. Some samples of student work and the misconceptions revealed in their work are included in Table 3.

Table 3. Misconceptions in tabular representations of data included in the story

Misconception: Data for speed and time were included instead of distance and time which is required for plotting distance-time graphs
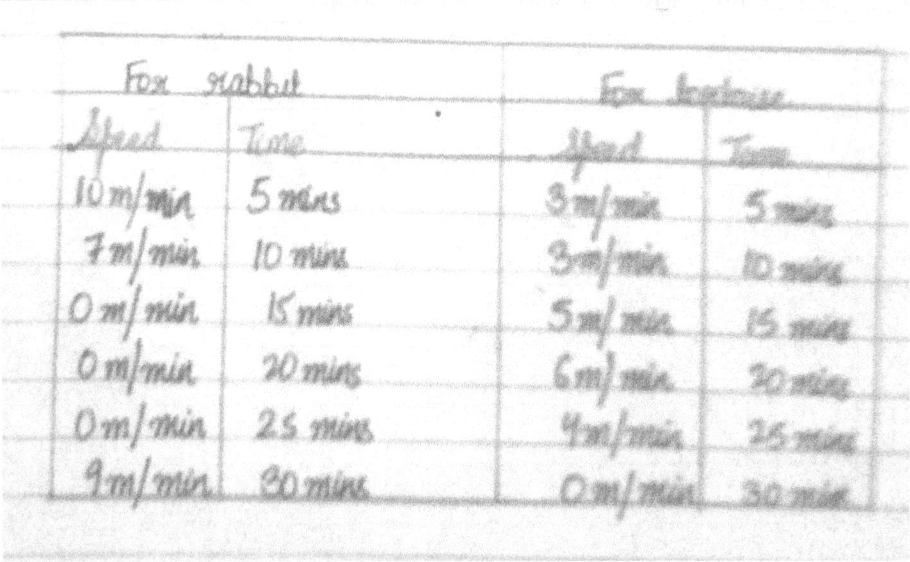

For rabbit		For tortoise	
Speed	Time	Speed	Time
10 m/min	5 mins	3 m/min	5 mins
7 m/min	10 mins	3 m/min	10 mins
0 m/min	15 mins	5 m/min	15 mins
0 m/min	20 mins	6 m/min	20 mins
0 m/min	25 mins	4 m/min	25 mins
9 m/min	30 mins	0 m/min	30 min

Several procedural errors emerged from students' *graphical representations*. It included distance-time graphs without any mention of the titles of the axes, units of the quantities being measured, use of different origins for two quantities being represented by the two axes. In a few cases, students did not provide index/ key that enables the reader to decode which character's motion is being represented by a plot line. Some students failed to choose the appropriate scale unit corresponding to their data sets or plotted incorrect coordinates in graphs. Representative student work samples are included in Table 4

Table 4. Misconceptions in plotting distance-time graphs

Misconceptions: • Use of two origins for representation of two physical quantities • Names of the physical quantities represented by the two axes are missing
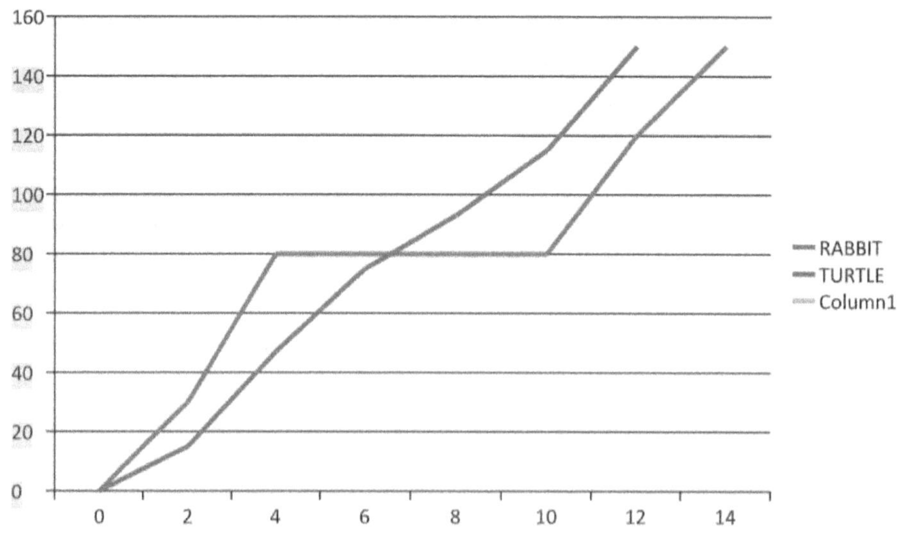

Explaining phenomena scientifically

The competency of evaluating and designing scientific enquiry is required to evaluate reports of scientific findings and investigations critically. It relies on the ability to recognise questions that can be investigated scientifically. This competency requires a knowledge of the key features of a scientific investigation, such as what things should be measured, what variables should be changed or controlled, and what action should be taken so that accurate and precise data can be collected. It requires an ability to evaluate the quality of data, which in turn depends on recognising that data are not always and completely accurate.

This task encouraged students to establish synchronisation between the verbal and visual descriptions of the story. Student work samples revealed that there existed mismatch between datasets or units of physical quantities used in the story and as presented in tabular and graphical representations of the story. Some students plotted hypothetical data which was not mentioned in their stories or included in their tabular representation. Table 5 presents commonly held student misconceptions that emerged from student work.

Table 5. Misconceptions in establishing synchronisation among story, tabular, and graphical representations

Misconceptions revealed

- Data points of speed were not provided in the story but were included in the table
- Graph is unintelligible

Story

This is a story of a girl named Sakshi. She lived at Dariyapur village, Haryana and studied at University of Delhi. On11th August, she started her journey to college at 6:10 am. Her Papa dropped her to bus stop as she was having a bus at 6:20am. The bus stop was 3km away from house. She hoped she will reach college on time. But unfortunately, the DTC bus suddenly stopped at Issapur after covering 20km at 6:40am. Then she had to wait there for 10 minutes.

At 6:50am another bus came and then she reached Najafgarh 90km away at 7:25am. But who knows there was another problem waiting there. Due to the festival of Raksha Bandhan there was a traffic jam in Najafgarh. The bus started decelerating and began crawling. Now the clock struck 7:35am. Thank God, the jam got over. Bus accelerated itself and she reached Dwarka Mor at 7:45am after covering 102km.

There before boarding the train, the security check ups took another 5 minutes. She boarded the train at 7:50am. There was a slight relief there but also a fear of getting late. After crossing 15 stations and covering 138 km she reached at Rajiv Chowk. There she changed for Yellow line at 8:25 am. She reached Visavidyalya station at 8:40 am which crossed the distance 150 km. After coming out of platform she saw a long queue at exit gate. But after politely requesting others she managed to get out of metro station at 8:45am. There she walked for another 100m to reach college. So, she reached college at 8:50am.

S .NO.	Distance covered(km)	Time taken(min)	Velocity(km\ min)
1.	3	10	0.3
2.	23	30	0.7
3.	23	40	0.5
4.	93	75	1.24
5.	93	85	1.09
6.	105	95	1.10
7.	105	100	1.05
8.	141	135	1.04
9.	153	150	1.02
10.	153	195	0.78
11.	153.100	200	0.76

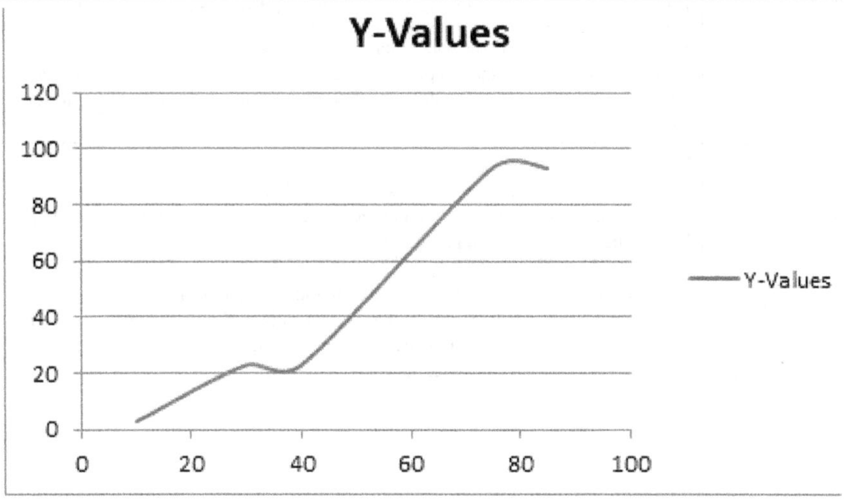

Formative Feedback

Hattie and Timperley (2007) in their seminal work observed that effective formative feedback entails guiding learners to evaluate their own work with respect to the learning goals. They emphasized that a teacher's job is to clarify the learning goals to make it accessible to learners, provide conducive learning environment, enable them to develop their unique learning strategies and error detection skills. Task-specific guidance can only be provided if teacher is well-aware of the learning gaps students are facing.

This formative assessment task provided clear window to students as well as the teacher about specific errors students were committing. She asked probing questions with reference to each of the four misconceptions elicited: writing stories, tabulating data included in the stories, drawing distance-time graphs, and establishing synchronisation between verbal and visual descriptions of the story as described in Table 6.

This formative feedback approach was targeted to "enhances a student's more general 'learning power' in terms of their motivation, confidence and strategic awareness of their own learning strengths and weaknesses" (Broadfoot, 2008, p. 124), consequently, making learners self-propelled learners.

Table 6. Formative feedback

Task	Prompts used by the teacher as formative feedback
Writing stories containing mathematical data	Do you require any form of mathematical data in this story? Give reasons to support your answer. Are the quantities provided to represent distance and time realistic? Is your story representing the data required to plot distance-time graph? Reflect.
Tabulating the data in the story	Does your table include the data required to plot distance-time graph? Which column of your table represent distance? Have you named it appropriately? Are the units you are using to describe time in the table same as the units in story? Check.
Plotting distance-time graph	Have you named the two axes? Which units are being used to represent time and distance? Can you name the ordered pair being represented by the plot points in your graph? Please check if the ordered pair are matching with the data in the table.
Establishing synchronisation between stories, table, and graph	Check if the data plotted in graph is there in the table and story. By reading graph, can you suggest the time taken by a character to travel a particular distance? Now, match it with the data provided in the table and story.

CONCLUSION

PISA framework on scientific literacy observes that in order to develop various 21st century skills among students, such as, critical thinking, communication, creativity and problem-solving, it is essential to use competency-based assessments which encourage students to apply their scientific literacy developed in classrooms to solve problems set in real-world situations (OECD, 2022).

This study used creative story writing as a competency-based formative assessment for assessing student understanding of graphs. The task comprised of several steps: writing creative stories involving motion of two characters, tabulating the data included in stories, plotting their verbal description visually in the forms of distance-time graphs. Clearly, it was crucial that students were able to make explicit connections between visual features demonstrated in their graph with their meaning encoded in the verbal descriptions (stories).

Underscoring the importance of assessing the process as well as product of formative assessment, this task illustrates varied student-held misconceptions that emerged at different steps as listed in Tables 2-5. Student-held misconceptions are indicative of the different ways in which they approach a task. Thereby, signalling that same feedback cannot work for all students. Teachers need to diversify their feedback mechanisms to make it more specific to students' areas of difficulty. A comprehensive list of probing questions used by the teacher for providing formative feedback is listed in Table 6.

Conclusively, this study suggests that all students are unique and adopt unique learning pathways to solve varied assessment tasks set in authentic contexts. Teacher, in such cases, gets an opportunity to support her learners by planning differentiated instruction and differentiated feedback. By this means, scientific literacy developed in classroom context becomes useful in real world situations as students learn appropriate ways of meaning-making and problem solving through carefully designed competency-based assessments.

REFERENCES

Albano, G., & Pierri, A. (2014). Digital storytelling for improving mathematical literacy. In *Proceedings of Problem@Web International Conference* (pp. 23-34). Academic Press.

Ashbrook, P. (2020). Becoming scientifically literate. *Science and Children.*

Balakrishnan, C. (2008). Teaching secondary school mathematics through storytelling [Unpublished thesis, Simon Fraser University].

Bansal, G. (2014). *Assessment, Curriculum and Pedagogy: Towards conceptualising an alternative framework of assessment*. [Thesis, University of Delhi: Delhi].

Bansal, G. (2017). School-based professional learning community: Empowering teachers as assessment leaders in the context of change. In In I.H. Amzat, and N.P. Valdez (Eds.) *Teacher Empowerment toward Professional Development & Practices: Perspectives Across Borders*. (Pp. 255-268), Springer: Singapore. doi:10. https://www.springer.com/gp/book/9789811041501 doi:.1007/978-10-4151-8

Bansal, G. (2020). Understanding gaps in teacher interpretation of formative assessment evidence. *The School Science Review*, *101*(377), 67–72.

Black, P., & Wiliam, D. (1998). Assessment and classroom learning. *Assessment in Education: Principles, Policy & Practice*, *5*(1), 7–74. doi:10.1080/0969595980050102

Broadfoot, P. (2008). *An Introduction to Assessment*. Continum.

Care, E. (2020). *Optimising assessment for all: Assessment as a stimulus for scaling 21st century skills in education systems*. The Brookings Institution.

Casey, B., Erkut, S., Ceder, I., & Young, J. M. (2008). Use of a storytelling context to improve girls' and boys' geometry skills in kindergarten. *Journal of Applied Developmental Psychology*, *29*(1), 29–48. doi:10.1016/j.appdev.2007.10.005

Central Board of Secondary Education (CBSE). (2021). *Understanding competency-based learning for scientific literacy*. CBSE. https://cbseacademic.nic.in/cbe/documents/SAS_CBLF_Scientific-Literacy-Final.pdf

Chen, Y.-C., Hand, B., & McDowell, L. (2013). The effects of writing-to-learn activities on elementary students' conceptual understanding: Learning about force and motion through writing to older peers. *Science Education*, *97*(5), 745–771. doi:10.1002ce.21067

Comfort, K. B., & Timms, M. (2018). A Twenty-First Century Skills Lens on the Common Core State Standards and Next Generation Science Standards. In E. Care, P. Griffin, & M. Wilson (Eds.), *Assessment and Teaching of 21st Century Skills: Research and Applications* (pp. 131–144). Springer. doi:10.1007/978-3-319-65368-6_8

Curcio, F. R. (1989). *Developing Graph Comprehension. Elementary and Middle School Activities*. National Council of Teachers of Mathematics.

DeBoer, G. E. (2000). Scientific literacy: Another look at its historical and contemporary meanings and its relationship to science education reform. *Journal of Research in Science Teaching*, *37*(6), 582–601. doi:10.1002/1098-2736(200008)37:6<582::AID-TEA5>3.0.CO;2-L

Fadel, C., Bialik, M., & Trilling, B. (2015). *Four-Dimensional Education: The Competencies Learners Need to Succeed*. Center for Curriculum Redesign.

Friel, S. N., Curcio, F. R., & Bright, G. W. (2001). Making Sense of Graphs: Critical Factors Influencing Comprehension and Instructional Implications. *Journal for Research in Mathematics Education*, *32*(2), 124–158. doi:10.2307/749671

Gallup. (2013). *21st century skills and the workplace; A 2013 Microsoft Partners in Learning and Pearson Foundation Study*. Gallup. https://www.gallup.com/poll/162818/21st-century-skills-linkedwork-success.aspx

Ginsburg, H., & Opper, S. (1969). *Piaget's Theory of Intellectual Development*. Prentice Hall.

Glazer, N. (2011). Challenges with Graph Interpretation: A Review of the Literature. *Studies in Science Education*, *47*(2), 183–210. doi:10.1080/03057267.2011.605307

Goral, M. B., & Gnadinger, C. M. (2006). Using storytelling to teach mathematical concepts. *Australian Primary Mathematics Classroom*, *11*(1), 4–8.

Hadzigeorgiou, Y. (2016). Imaginative Science Education. Springer. doi:10.1007/978-3-319-29526-8

Hattie, J., & Timperley, H. (2007). The power of feedback. *Review of Educational Research*, *77*(1), 81–112. doi:10.3102/003465430298487

Heritage, M. (2010). *Formative Assessment: Making It Happen in the Classroom*. Corwin. doi:10.4135/9781452219493

Jones, E. A., Voorhees, R. A., & Paulson, K. (2002). *Defining and Assessing Learning: Exploring Competency-Based Initiatives*. Council of the National Postsecondary Education Cooperative Working Group on Competency-Based Initiatives.

Keys, C. W., Hand, B., Prain, V., & Collins, S. (1999). Using the science writing heuristic as a tool for learning from laboratory investigations in secondary science. *Journal of Research in Science Teaching*, *36*(10), 1065–1084. doi:10.1002/(SICI)1098-2736(199912)36:10<1065::AID-TEA2>3.0.CO;2-I

Kim, H., & Care, E. (2020). *Optimizing Assessment for All: Classroom-based Assessments of 21st century Skills in the Democratic Republic of Congo, The Gambia, and Zambia*. The Brookings Institution.

Lacefield, W., & Markert, L. (2019). Creative Writing in the Mathematics Classroom. *Proceedings of the Annual Meeting of* the Georgia Association of Mathematics Teacher Educators, *13*(1). 10.20429/gamte.2019.130106

Leikin, R. (2009). Exploring mathematical creativity using multiple solution tasks. In R. Leikin, A. Berman, & B. Koichu (Eds.), *Creativity in mathematics and the education of gifted students* (pp. 129–145). Sense Publishers. doi:10.1163/9789087909352_010

Lemonidis, C., & Kaiafa, I. (2019). The Effect of Using Storytelling Strategy on Students' Performance in Fractions. *Journal of Education and Learning*, *8*(2), 165–175. doi:10.5539/jel.v8n2p165

McKinsey Global Institute. (2017). *Jobs lost, jobs gained: What the future of work will mean for jobs, skills, and wages*. McKinsey. https://www.mckinsey.com/featured-insights/future-ofwork/jobs-lost-jobs-gained-what-the-future-of-work-will-mean-for-jobs-skills-and

National Research Council. (2012). *A framework for K-12 science education: Practices, crosscutting concepts, and core ideas*. The National Academies.

NCERT. (2006). *Position Paper National Focus Group on Examination Reforms*. NCERT.

NGSS Lead States. (2013). *Next generation science standards: For states, by states*. The National Academies.

OECD. (2005). *Formative Assessment: Improving Learning in Secondary Classsrooms*. OECD.

OECD. (2017). Future of Education and Skills 2030: Reflections on Transformative Competencies 2030 (No. EDU/EDPC(2017)16/ANN5). Paris, France: Organization for Economic Cooperation and Development.

OECD. (2019). *PISA 2018 Assessment and Analytical Framework, PISA*. OECD Publishing. doi:10.1787/b25efab8-

Partnership for 21st Century Skills (P21). (2009). *Framework for 21st century learning*. Tucson. http://www.p21.org/storage/documents/P21_Framework.pdf

Roegiers, X. (2016). *A Conceptual Framework for Competencies Assessment: Current and Critical Issues in the Curriculum and Learning*. UNESCO: IBE.

Saavedra, A., & Opfer, D. (2012). Learning 21st—century skills requires 21st—century teaching. *Phi Delta Kappan*, *94*(2), 8–13. doi:10.1177/003172171209400203

Schiro, M., & Lawson, D. (2004). *Oral storytelling and teaching mathematics: Pedagogical and multicultural perspectives*. Sage Publications. doi:10.4135/9781452229751

Schleicher, A. (2012). Preparing teachers and developing school leaders for the 21st century: Lessons from around the world. OECD Publishing.

Scoular, C., Ramalingam, D., Duckworth, D., & Heard, J. (2020). *Assessment of general capabilities: Skills for the 21st-century learner. Final report.* Australian Council for Educational Research. https://research.acer.edu.au/ar_misc/47

Shah, P., & Carpenter, P. A. (1995). Conceptual Limitations in Comprehending Line Graphs. *Journal of Experimental Psychology, 124*(1), 43–61. doi:10.1037/0096-3445.124.1.43 PMID:7844506

Shah, P., & Hoeffner, J. (2002). Review of Graph Comprehension Research: Implications for Practice. *Educational Psychology Review, 14*(1), 47–69. doi:10.1023/A:1013180410169

Trilling, B., & Fadel, C. (2012). A game of skills. What talents do young people need to thrive in the 21st century? *Royal society for the encouragement of arts, manufactures and commerce journal.*

Turiman, P., Omar, J., Daud, A. M., & Osman, K. (2012). Fostering the 21st century skills through scientific literacy and science process skills. *Procedia: Social and Behavioral Sciences, 59*, 110–116. doi:10.1016/j.sbspro.2012.09.253

UNICEF. (2019). *More than half of South Asian youth are not on track to have the education and skills necessary for employment in 2030.* UNICEF. https://www.unicef.org/press-releases/more-half-south-asian-youth-are-not-trackhave-education-and-skills-necessary

Wallace, C., Hand, B., & Prain, V. (2004). *Writing and learning in the science classroom.* Kluwer. doi:10.1007/978-1-4020-2018-6

Wiliam, D., & Thompson, M. (2007). Integrating assessment with instruction: what will it take to make it work? In C. A. Dwyer (Ed.), *The Future of Assessment: Shaping Teaching and Learning* (pp. 53–82). Lawrence Erlbaum Associates.

World Economic Forum. (2018). *The Future of Jobs Report.* WEF. https://www3.weforum.org/docs/WEF_Future_of_Jobs_2018.pdf

Zevenbergen, R., Dole, S., & Wright, R. J. (2004). *Teaching Mathematics in Primary Schools.* Allen & Unwin.

Zucker, A. (2021). Teaching scientific literacy. *Science Teacher (Normal, Ill.).*

Chapter 4
Computational Thinking in Science Teaching and Learning

Ayodele Abosede Ogegbo
University of Johannesburg, South Africa

Umesh Ramnarain
University of Johannesburg, South Africa

ABSTRACT

Given the importance of inquiry in science education, various computational tools and emerging 4IR technologies have been embraced in recent years to improve how inquiry is carried out across science disciplines. However, there is growing concern about how schools can better guide students and teachers in leveraging the technological affordances of the 4IR. In response to this, the authors propose that one pertinent approach to the development and preparation of science teachers for the future of 4IR classrooms is to teach them how to think computationally. Based on a recent literature review, this chapter offers insight into how computational thinking can be used as a pedagogical approach to science teaching in a 4IR classroom.

INTRODUCTION

Schwab (2016) described the fourth industrial revolution (4IR) as a technological era characterised by the fusion of modern knowledge and skills in unprecedented ways to foster and broaden innovative collaboration between humans, the digital, biological and physical worlds. Considering the exponential rate at which the use of scientific and technological innovations such as robots, artificial intelligence, 3D printing, the internet of things, augmented reality, virtual reality, mixed reality, and

DOI: 10.4018/978-1-6684-6932-3.ch004

many other advances and innovative systems are becoming imperative to economic development and globalisation, there is a growing trend of interest on the set of skills, training, expertise, abilities and education that learners require in order to succeed in the changing world of work brought about by 4IR. Research claims that one approach to teaching the essential skills needed for the 4IR is to teach computational thinking (National Research Council (NRC, 2010; Wesse, 2017; Balanskat et al., 2018). Berland and Wilensky (2015) defined computational thinking (CT) as "the ability to think with the computer-as-tool" (p. 630). However, Wing (2006) stated that CT employs computers as a thinking tool, universal mindset, and skill set. She claims CT "draws on the fundamental ideas of computer science to account for solving problems, designing systems, and understanding human behaviour" (p.33). More specifically, CT includes breaking down a complex task into smaller and more manageable components (decomposition), identifying and defining trends or patterns within a problem (pattern recognition), identification of particular similarities and differences between similar problems to work towards a solution (abstraction), development of step by step guidelines for solving a problem and can be used again to answer similar problems (algorithm) and use of technological tools to mechanise problem solutions (automation) (ISTE & CSTA, 2011; Selby, 2013; Yadav et al., 2018). This suggests that components of computational thinking are made up of computational concepts, principles, methods, languages, models, and tools that are commonly found in the study of computer science. Hence, computational thinking may include the following:

"reformulation of difficult problems by reduction and transformation; approximate solutions; parallel processing type checking and model checking as generalisations of dimensional analysis; problem abstraction and decomposition; problem representation; modularisation; error prevention, testing, debugging, recovery, and correction; damage containment; simulation; heuristic reasoning; planning, learning, and scheduling in the presence of uncertainty; search strategies; analysis of the computational complexity of algorithms and processes; and balancing computational costs against other design criteria. Concepts from computer science such as algorithm, process, state machine, task specification, formal correctness of solutions, machine learning, recursion, pipelining, and optimisation also find broad applicability" (NRC, 2010, p.3).

Wing (2008) claims that mastering computational thinking skills and related practices is vital to understanding how to process information about problems and human behaviours using computing methods. In light of this, computational thinking can be viewed as a powerful approach for teaching students how to think critically and solve problems. According to the reports from the 2010 workshop on the scope and nature of computational thinking, promoting computational skills in K-12 education would provide students with a way to think about computing, help students sustain the rising interest in computing as a profession, increase equitable access to modern society's resources, give students the ability to navigate a technological society more effectively, and allow students to more effectively adopt, use, and develop computational tools (NRC, 2010). Thus, transition into the 4IR requires that individuals possess the analytical ability to create and use digital tools in solving complex environmental problems and the skills and knowledge required to thrive in the 4IR workspace. In the wake of the 4IR, researchers have argued that computational thinking should be on equal footing with the essential skill of reading, writing, and arithmetic being taught in schools to help develop students' mathematical and scientific literacy (Grover, 2018; Wing, 2006). The need for everyone to be proficient in computational thinking (CT) in an information-based society has increased the focus on integrating CT into K–12 education and expanding it outside of computer science courses (Yadav et al., 2014). As a result, CT is increasingly being incorporated into the national curriculum in many countries as part of a subject or topic taught in schools or throughout the curriculum (Balanskat et al., 2018). It is believed that developing learners' thought processes and CT skills could increase their enrolment and achievement in science education globally. Hence, this chapter aims to provide helpful insight into how teachers can incorporate computational thinking into science instructions and a perspective on barriers to integrating CT in science classrooms and possible ways of addressing them was to integrate three dimensions (directionality, openness, and process) into the design of practical assessment tasks; and the other was to assess the three dimensions of CT comprehensively including computational concepts, practices, and perspectives.

COMPUTATIONAL THINKING AND SCIENCE EDUCATION

According to the Next Generation Science Standards (NGSS) and the Framework for K-12 science education in the United States, one of the primary goals of science education is to cultivate students' scientific habits of mind, teach students how to reason in a scientific context, and develop students' capability to engage in scientific and engineering practices as they prepare for their individual and professional lives in a technologically rich and scientifically complex environment (NGSS Lead States, 2013). The council of state science supervisors in the United States (n.d) claims that students who actively engage in scientific and engineering practices tend to better understand crosscutting concepts and disciplinary ideas. In light of this, the NGSS framework identified mathematics and computational thinking as one of the eight science and engineering practices students should use in conjunction with disciplinary core ideas and crosscutting concepts to understand phenomena and find solutions to problems (NGSS Lead States, 2013). Computational thinking is defined in the framework as the ability to use computational tools, which include computers, and computational methods like building simulations, statistically analysing data, applying quantitative relationships, and mathematically testing design solutions as problem-solving tools (NRC, 2012). The NGSS emphasises the importance of improving science content through effective computational thinking strategies such as "organising and searching data, creating sequences of steps called algorithms and using and developing new simulations of natural and designed systems" (Jenner, n.d, p.13). Educational technology advancements have also improved the computational nature of STEM fields as applied in the workplace. As a result, research has shown that incorporating CT into science content and context provides learners with a more realistic portrayal of how science is practised in the workplace and improves learners' higher-order thinking (Swanson et al., 2019).

Similarly, incorporating computational thinking into the science classroom through the use of computational tools such as simulations, agent-based models, and large data sets coupled with appropriate data analysis strategies can allow students to conduct a broader range of investigations, broaden students experiences and learning, and assist students in visualising the emergent properties of systems (Mcgowan et al., 2019; Swanson et al., 2019). Furthermore, incorporating CT into science instruction allows students to make sense of phenomena related to core disciplinary ideas and crosscutting concepts (Miller et al., 2020). This implies that the intentional application of computational tools and skillsets can enhance the learning of STEM content. However, the appropriate progression for computational thinking depends on the intentional integration of CT concepts and practices with specific scientific ideas (Miller et al., 2020). This implies that the teaching and assessment of CT are highly dependent on the context within which the teaching and assessment are performed.

FRAMEWORK FOR INTEGRATING COMPUTATIONAL THINKING

Numerous frameworks for adopting computational thinking have been proposed, which can assist teachers in organising and incorporating indicators of CT capabilities in their lesson activities while improving learners' CT skills. For instance, Brennan and Resnick (2012) described the three "dimensions" of CT to include computational concepts (e.g., sequences, loops, parallelism, events, conditionals, operators, and data), computational perspectives (including evolving views of self and the relationships with others and with their digital surroundings, manifested through expressing, connecting, and questioning), and computational practices (e.g., incremental and iterative manner, testing and debugging, reusing and remixing, and abstracting and modularising). Thus, these components cover the what, how, and why of CT. Following Brennan and Resnick's (2012) framework, Zhong et al. (2016) designed the Three-Dimensional Integrated Assessment (TDIA) framework to thoroughly assess the three dimensions of CT (concepts, practices and perspectives) and integrate three dimensions (directionality, openness, and process) into the design of computational assessment tasks. The framework uses scaffolding methods and three types of assessment tasks to follow students' performance.

On the other hand, Weintrop et al. (2016) proposed a "CT taxonomy framework" that can be used to incorporate CT into Science, Technology, Engineering and Mathematics (STEM) disciplines. The taxonomy framework comprises twenty-two CT practices grouped into four major categories: data practices, modelling and simulation practices, computational problem-solving practices, and systems thinking practices. Each category is made up of five to seven different practices. However, Kotsopoulos et al. (2017) proposed a CT pedagogical framework that is particularly effective for students and teachers to experience CT fully. The framework identified four types of instructional experiences: unplugged (engaging in CT activities without the use of computers), tinkering (entails taking things apart and making adjustments to existing objects), making (entails activities in which the primary focus is on the creation of new objects), and remixing (involves the appropriation of items or components of objects for use in other objects or other purposes). These instructional experiences provide insight into pedagogical strategies and resources teachers can use to improve CT learning experiences. Moreover, literature identified pedagogical strategies such as problem-solving, open-ended tasks, project-based learning, group work and collaboration, inquiry-based learning, challenge-based learning, blended learning, flexible learning, and peer-to-peer learning as student-centred pedagogical approaches frequently used by teachers (Bower et al., 2017; TuckerRaymond et al., 2021; Ogegbo & Ramnarain, 2021).

Similarly, Grover and Pea (2018) developed a competency framework to assist instructors and designers in developing a curriculum to improve students' CT skills. The framework consisted of two core competencies: CT concepts and CT practices. The six sub-competencies for "CT concepts (are logic and logical thinking, algorithms and algorithmic thinking, pattern and pattern recognition, abstraction and generalisation, and evaluation and automation). Whereas the five sub-competencies for CT practices include (problem decomposition, creating computational artefacts, testing and debugging, iterative refinement, and collaboration and creativity)" (Grover & Pea, 2018, p.23).

Due to the lack of consensus on the definition of computational thinking, Fraillon et al. (2019) synthesised the definition of CT as "an individual's ability to recognise aspects of real-world problems which are appropriate for the computational formulation, and to evaluate and develop algorithmic solutions to those problems so that the solutions could be operationalised with a computer "(p27). To further enhance how CT can be assessed among learners, Fraillon and his colleagues proposed a CT framework comprising two strands of problem that include conceptualisation and operationalising solutions. According to the authors, problem conceptualisation recognises that generating solutions requires that problems must first be adequately understood and structured so that algorithmic or systems thinking may assist in producing solutions. This strand consists of three components: knowledge and understanding of digital systems, issue formulation and analysis of problems, and data collection and representation. At the same time, operationalising solution involves developing, implementing, and assessing computer-based system responses to real-world problems. It encompasses the iterative processes of developing, executing, analysing, and assessing algorithmic solutions (as prospective programming bases) to real-world issues. This strand comprises an understanding of user needs and how they will interact with the technology under development. This strand is divided into two parts: solution planning and evaluation, and algorithm, program, and interface development (Fraillon et al., 2019). A comprehensive review of the various frameworks reveals that CT is a problem-solving strategy with distinct cognitive and non-cognitive components within an integrated framework of CT capabilities. The frameworks reviewed in this section provide a broad view of how CT can be implemented in science classrooms.

METHODS FOR INCORPORATING COMPUTATIONAL THINKING INTO SCIENCE EDUCATION

Policymakers across the globe recognise the importance of fostering key competencies and abilities in the teaching process to enable students to thrive in the 4IR workplace. However, the numerous discussions on preparing students for 4IR have not given teachers the tools to help students express these goals in their learning. Though educational policies and reforms prioritise enhancing student learning, teachers play a crucial role in advancing high-quality science instruction through classroom activities. In order to cultivate an experiential and systematic approach that expands the capabilities of computational thinking practices in science education, teachers should be able to take advantage of the technological affordances of the 4IR. Research has shown that learners can obtain competencies and skills such as problem-solving, computational thinking, critical thinking, collaboration, communication, creativity and innovation needed for the future workplace through game-based and inquiry-based learning (Kuhlthau et al., 2015; Weese, 2017). This implies that computational thinking can be taught through inquiry-based learning. For instance, in a study that described how preservice teachers integrate computational thinking in physical science classrooms, Tsakeni (2021) stated that teachers used CT as a problem-solving strategy to facilitate inquiry-based practical work in deprived classrooms. Teachers integrated computational thinking into their science lessons by engaging students in virtual hands-on activities through an online laboratory (Praxi Labs) that provides immersive and interactive 3D-simulation experiences of a real laboratory for learners, PhET simulation application, CK-12 app, and EveryCircuit app (Tsakeni, 2021). The apps were assessed on the school's computer lab, learners' smartphones and tablets. The author also discovered that teachers employed computational thinking techniques like problem decomposition, pattern recognition, abstraction, and algorithm when using the PhET simulation application to conduct virtual hands-on activities to investigate the relationship between speed and force. According to the author, using online labs and other applications allowed students to engage in trial-and-error and improvement exercises, build circuits, manipulate the materials and variables they want to use, take precise measurements, record results, and draw conclusions with a single click. Teachers used these applications with excel sheets to enable learners to visualise their results (data representation) in the form of graphs. The author concluded that using computational thinking enabled the preservice science teachers to solve challenges that inhibit the successful implementation of inquiry-based practical work.

Similarly, Tucker-Raymond et al.'s (2021) study within US schools showed that middle school science teachers integrated computational thinking (CT) into teaching climate systems and climate change through computer game design. Due to teachers' lack of experience with computational thinking practices, they shifted their approach to teaching CT in science lessons from teacher-led to distributed expertise (student-centred). As a result, students were required to struggle through the problem independently, relying on each other and other resources during their initial attempt to integrate computational problem-solving. In addition, students were required to be as knowledgeable as or even more than their teachers. Distributed expertise assists teachers in organising student-centred learning strategies that foster student agency by fostering a participatory culture in the classroom and incorporating complex computational tasks (Lye & Koh, 2018). To effectively integrate CT and support students as computational problem solvers using distributed expertise, students were given autonomy over the climate issues they addressed and the game design they chose to represent their ideas. Teachers also had to co-learn with students using the Scratch platform and figure out how to help students solve problems that arose. Tucker-Raymond et al. (2021) observed that students built interdependence among each other and developed confidence in their ability to create computer games about climate change using scratch features and CT practices or strategies such as computational problem-solving, debugging, and programming (remixing existing code). This implies that the integration of CT skills can also be fostered through game-based learning. A crucial aspect of using CT with Scratch is the ability to visualise the phenomenon, which enhances conceptual understanding of a topic (Shodiev, 2015). Tucker-Raymond and his colleagues concluded that even though teachers lacked the skills and knowledge necessary to incorporate computational thinking practices into science lessons effectively, they viewed the shift to distributed expertise as a way to put those skills into practice and support student-centred learning. The result of the study demonstrated the value of leveraging distributed systems of expertise among students through online search strategies, curricular resources, and publicly identifying student experts to peer as a promising strategy for implementing CT in science classrooms.

Furthermore, Shodiev (2015) indicated that students could experiment with scratch codes to create parabolic trajectories and use the scratch platform to simulate projectile motion in physics and other physical phenomena in our immediate surroundings, such as the motion of colliding spheres and momentum conservation. Moreso, students can build the simulation with or without algebraic tools depending on their math proficiency. This simulation allows students to improve their CT by tinkering and concentrating on physics ideas themselves (Shodiev, 2015). Moreso, research indicates that computational platforms (Web-based simulation authoring tools) like Agentsheets and Agent cubes, robotics kits, and tangible media like Arduino and Gogo Boards can be used in classrooms to improve learners' involvement with the various component components of CT (Grove & Pea, 2013; Ogegbo & Ramnarain, 2021).

Though the use of computational tools has been found to promote students' deeper learning of STEM subjects, researchers have also argued for the inclusion of CT in science classrooms through unplugged activities because not every science classroom has access to technology. For instance, Miller et al. (2020) investigated how computational thinking is used to teach science in a fifth-grade elementary classroom in the United States. The authors designed multiple literacies in project-based learning units and utilised several elements of computational thinking concepts and practices described by Grover and Pea (2013, 2018) to enhance students' disciplinary core idea of the Particle Nature of Matter and crosscutting concept of Patterns. Students were given a task that required them to revise or iterate (a computational thinking practice) how to break up the way they tasted sweetness into categories. After determining the categories, students were asked to decompose (a computational thinking practice) the phenomenon problem into smaller components based on pattern finding (a computational thinking concept). Miller et al. (2020) discovered that students could categorise the foods into three groups, effectively finding patterns (identifying recurring elements) in the similarities and differences between the foods. As students revised and iterated how to sort tastes into various categories, they re-engaged with the concept of matter being made up of particles. The authors indicated that the use of patterns encouraged an iteration of ideas around the phenomenon of taste and discussion about how the phenomenon has patterns that may be used to distinguish between food samples with similar and distinct compositions. Students were observed using computational thinking to iteratively develop a brief explanation of the phenomenon of taste, which is the chemical reaction to various particle compositions that make up foods. According to the authors, the Computational thinking practices of decomposition and iteration enabled the students to rethink the phenomenon of taste and smell as a phenomenon with several phases. Thus improving their understanding of the particle nature of matter. The authors concluded that students were able to recognise that there are some common compositions of particles in similar-tasting foods, thus arriving at a deeper understanding of the particle nature of matter through engaging in the decomposition of the phenomenon into smaller parts.

Similarly, Peel et al.'s (2022) study on the integration of computational thinking in a science lesson demonstrated how teachers employed various algorithm explanation activities to promote students' learning about the process of natural selection. Findings of the study showed that using unplugged algorithmic explanation as an instructional strategy helps to enhance students' computational literacy and fosters the development of students' scientific ideas and algorithmic logic and practices. Based on their findings, Peel et al. (2022) suggested that using unplugged activities can assist students in acquiring CT in more profound ways and allow them to use CT more successfully in different circumstances based on their findings.

ASSESSING COMPUTATIONAL THINKING

Assessment is a crucial component of the learning process (Stegeman et al., 2016). Assessment directs student learning, aids teachers in evaluating the degree to which learning objectives are being reached, and gives the teacher and student feedback (Ihantola et al., 2010). The significance of assessing CT practices has been explicitly emphasised in light of the increased interest in using computational thinking in educational contexts (Grover & Pea, 2013; Zhong et al., 2016; Weintrop et al., 2021). However, given the wide range of techniques used for teaching CT, various facets of CT, and the various educational contexts in which students learn CT, there is no agreement or standardisation on how computational thinking should be assessed (Grover & Pea, 2013; Weintrop et al., 2021). Nonetheless, research contends that when grading complex skills like CT, the structure, as well as the levels of the construct, must be considered because not all learning objectives can be assessed using the same tasks or assessment approach (Seufert et al., 2021; Weintrop et al., 2021).

A literature review study conducted by Tang et al. (2020) reported that CT could be assessed through selected-response and/or constructed-response tests, multiple-choice questions and open-ended questions, performance or portfolio assessment, pre-and post-tests, questionnaires and interviews. However, the use of performance assessment in measuring CT appears to provide a variety of advantages over traditional assessments due to its capacity to examine higher-order thinking. Moreover, Tang et al. (2020) stated that many existing computational thinking assessments lack valid and reliable inferences about student learning. Due to a lack of consensus on a guiding framework, definition, or collection of proxies for assessing CT, there seems to be a problem with the quality of CT assessments in education. In light of this, Clarke-Midura et al. (2021) provide evidence on how to assess computational thinking using the Evidence Centered Design (ECD) approach. The authors pointed out that the ECD design approach allows teachers to establish meaningful conclusions regarding the learning of computational thinking based on repeatedly defining and assessing children's computational thinking knowledge, skills, and abilities. Furthermore, research indicates that computational thinking can also be assessed by involving students in closed, semi-open and open tasks while measuring their performance based on a set of rubrics/checklist that represents various CT skills and practices (Clarke-Midura et al., 2021; Zhong et al., 2016).

For learning to be effective, students must be aware of their level of performance on a task, how it compares to good performance, and what they can do to decrease the gap between those two (da Cruz Alves et al., 2019). This means that CT practices can be assessed in summative and formative ways. However, it is contended that using summative assessments to measure CT might not be beneficial for students who are classified as "low performers" because this type of assessment can cause them to lose confidence and believe in "fixed intelligence" and "inborn ability" rather than the possibility that effort can help them improve their CT skill and dispositions (Yin et at al., 2020). Furthermore, Yin et al. (2020) indicate that using formative assessment techniques such as planned-for interaction formative assessment and formal-and-embedded-in curriculum formative assessment may help enhance students' computational thinking skills and disposition. According to the authors, "planned-for interaction formative assessment is used during instruction but prepared deliberately before class to align closely with instructional goals; while formal-and-embedded-in curriculum formative assessment is designed to be implemented after a curriculum unit to ensure that students have reached important goals before moving to the next unit" (Yin et at al., 2020, p195). Furthermore, computational thinking concepts, dispositions, and related practices can be assessed using real-time data mining tools that retrieve and record students' activities, skill transfer tools that assess how well students can transfer their CT skills to different types of problems, contexts, and situations, perceptions-attitude scales, and CT vocabulary assessment (Román-González et al., 2019).

POSSIBLE BARRIERS AND CHALLENGES TO IMPLEMENTING COMPUTATIONAL THINKING IN SCIENCE CLASSROOMS

The growing interest in using computational tools in science education is related to their ability to promote learners' problem-solving competencies (ISTE & CSTA, 2011) and help students understand concepts underlying scientific phenomena and practices (Tsakeni, 2021). Incorporating computational thinking into science lessons requires teachers to change their classroom practices from a teacher-centred to a student-centred approach because computational thinking requires students to actively participate in practices and process skills rather than receiving direct instruction (Grover & Pea, 2013). Despite growing interest in using computational thinking in education, some argue that incorporating computational thinking principles into existing science curricula may be difficult for teachers and students (Grover & Pea, 2013; Weintrop et al., 2016). This may be because there is a lack of guidance and support available for science teachers to combine computational thinking with existing topics (Grover & Pea, 2013; Weintrop et al., 2016). In light of this, research claims that teaching computational thinking can present unique and sometimes complex challenges that many science teachers are unprepared to face (Saidin et al., 2021). These challenges are categorised as teacher-related challenges and student-related challenges.

Teacher-related challenges: These are personal and professional characteristics of teachers that influence their ability to integrate computational thinking in science classrooms effectively. They include factors like teachers' knowledge, beliefs, attitude, confidence and skills. As the integration of computational thinking has become an essential science practice, teachers need to have a sufficient understanding of computational thinking and its related practice to integrate it into their science lessons effectively. However, it seems that many teachers do not have a clear understanding of what computational thinking entails. For example, a quantitative study by Sands et al. (2018) investigated teachers' perceptions of computational thinking in K-12 classrooms. The study revealed that while teachers agreed with statements that conceptualised computational thinking as problem-solving, logical thinking, and algorithmic thinking, they also viewed computational thinking as mathematics, using a computer, and playing online games. While supporting the above finding, Hennico et al. (2021) reiterated that teachers have a widespread misconception regarding computational thinking, thus confusing computational thinking for programming or technology use and not applicable to science subjects. This assertion is not consistent with research conducted by Li et al. (2020), which shows that computational thinking is related to the cognitive process of problem-solving and that with computational thinking skills, people may understand the difficulty they are facing and produce a viable solution. Furthermore, it is argued that teachers' ideas of computational thinking tend to stem from misunderstandings about describing and recognising computational thinking in their instructional practices, making the integration of computational thinking complex (Cabrera, 2019).

Similarly, Corradini et al.'s (2017) study on primary school teachers in Italy revealed that most teachers did not comprehensively understand computational thinking. This lack of understanding has resulted in numerous issues for teachers, including time constraints for integrating computational thinking practices in science activities, low level of computer literacy, low self-efficacy in integrating CT, lack of confidence in developing computational thinking competencies in students, a lack of instructional materials, teachers' beliefs about students abilities and academic preparation, assessment techniques and adequate pedagogical knowledge to effectively teach science using computational thinking, and the need for infrastructure (Corradini et al., 2017; Kite &Park, 2020; Sand et al., 2018). This lack of knowledge could be attributed to most of them receiving little or no training in computational thinking. While agreeing with the previous finding, Peters-Burton et al. (2021) emphasised that unfamiliarity with computational thinking practices and related ideas could lead to a lack of confidence in labelling a lesson component with an incongruent computational thinking practice. Bower and Falkner (2015) contend that familiarity with computational thinking vocabulary does not always imply a profound understanding of computational thinking. In addition, the individual capacities of teachers to implement CT could also be impacted by their low level of knowledge (Rahayu & Osman, 2019). However, Hennico et al. (2021) indicate that teachers are more likely to implement computational thinking practices if they feel prepared and confident. In another survey study of 123 secondary science teachers, Kite and Park (2020) found that 24% of teachers who had heard the word CT but had not received CT professional development had accurate conceptions of CT, which is in line with Fraillons' (2019) conceptualisations of CT. However, teachers reported the following as perceived barriers to the integration of CT into science lessons: *I do not understand CT (35.34%), My students are not academically prepared for CT-infused lessons (15.26%), I do not understand how CT aligns with my existing science content (14.46%), I do not have access to necessary technology (11.65%), I do not have room in my curriculum (10.84%), I am not comfortable with technology (5.62%), and lack of administration support (2.81%).* These obstacles show that teachers require assistance in integrating CT with science instructions. Thus, it is vital to provide educators with the necessary support, and professional development that exposes them to computational thinking experiences, changes their perceptions about computational thinking and equips them to teach computational thinking in their classrooms. In light of this, Rahavu and Osman (2019) also emphasised the importance of providing teacher candidates with knowledge and awareness of computational thinking abilities in order for them to be ready and confident in tackling the difficulties of 4IR learning. This will help teachers to demonstrate a high level of knowledge and readiness for CT and align CT with their students' preparation as well as the subject and existing curriculum (Bower et al., 2017; Kite & Park, 2020).

Another significant observation is that student-related obstacles influence the implementation of computational thinking in science classrooms. For instance, in a study that explored influencing factors on college students' computational thinking skills through flipped-classroom instruction, Gong et al. (2020) found that student-to-student collaboration, learning motivation and learning strategy directly impacted students' learning of computational thinking skills. Stewart et al. (2021) added that psychometric properties such as student learning preference on perceived competence, intrinsic motivation and self-confidence also need to be considered in order to foster the learning of students' computational thinking skills. In addition, a quantitative study by Guggemos (2021) shows that student characteristics, home environment and learning opportunities play an important role in developing high school students' computational thinking skills. According to Guggemos (2021), the student characteristics include variables like gender, reasoning skills, mathematical skills, language skills, ability to program, computer literacy, computational thinking self-concept and self-determined motivation. While the home environment includes variables like parents' socioeconomic status and migration background, and learning opportunities include variables like formal learning opportunities during the school year, computer science instruction and duration of computer use.

Similarly, findings from the International Computer and Information Literacy Study (ICILS) conducted in 2018 reveal that socioeconomic background denoted by parental occupation, parental education, the number of books in the home, immigration and language background, access to computers, and experience using computers, all play a significant role in understanding the development and level of students' computational thinking skills (Fraillon et al., 2019). In addition, research claims that the integration of CT in science classrooms is hindered by disparities in access to a computer and other technological tools in school systems worldwide (Peel et l., 2022). Even though many schools have computer laboratories, most science classrooms still lack computers for all students (Hohlfeld et al., 2017). However, using unplugged CT techniques in science subjects can aid in minimising problems associated with technology access while still providing students with meaningful CT learning experiences.

CONCLUSION

Teaching CT in the context of science fosters access to powerful ways of thinking and valuable skills for various occupations and provides students with a more realistic view of science as practised today. This chapter provided an overview of the strategies and resources that can be used to integrate and assess computational thinking as a pedagogical approach to science teaching and learning in a 4IR classroom. The research report shows that computational thinking can be used as a pedagogical approach to science teaching through tools like simulation applications, virtual laboratories, scratch platforms, and strategies such as inquiry-based learning, game-based design, and distributed expertise. However, a few issues must be addressed before computational thinking can be effectively integrated as a pedagogical approach to science instruction in a 4IR classroom. First, technical assistance on how to use innovative technologies should be provided to students and teachers. Second, teachers should be supported with ongoing training on integrating CT into existing science content. Teacher education programs should expose trainee teachers to the use of computing tools in the science discipline. This will solve several issues, such as students' and teachers' lack of knowledge and confidence (Corradini et al., 2017; Saidin et al., 2021; Sand et al., 2018). To improve the incorporation of computational thinking into science classrooms, Lee et al. (2011) suggest using Use-Modify-Create (UMC) learning progression within computational environments. The use of UMC can help teachers improve CT acquisition while reducing cognitive stress by utilising curriculum resources with a scaffolded intensity of interaction.

REFERENCES

Balanskat, A., Engelhardt, K., & Licht, A. H. (2018). *Strategies to include computational thinking in school curricula in Norway and Sweden- European Schoolnet's 2018 Study Visit*. European Schoolnet.

Berland, M., & Wilensky, U. (2015). Comparing virtual and physical robotics environments for supporting complex systems and computational thinking. *Journal of Science Education and Technology*, *24*(5), 628–647. doi:10.100710956-015-9552-x

Bower, M., & Falkner, K. (2015). Computational Thinking, the Notional Machine, Pre-service Teachers, and Research Opportunities. In. D'Souza, D and Falkner, K. (Eds.), *Proceedings of the 17th Australasian Computing Education Conference Sydney, Australia*. CRPIT.

Bower, M., Wood, L. N., Lai, J. W., Highfield, K., Veal, J., Howe, C., Lister, R., & Mason, R. (2017). Improving the computational thinking pedagogical capabilities of school teachers. *The Australian Journal of Teacher Education*, *42*(3), 53–72. doi:10.14221/ajte.2017v42n3.4

Brennan, K., & Resnick, M. (2012). *New frameworks for studying and assessing the development of computational thinking*. Paper presented at the Proceedings of the 2012 Annual Meeting of the American Educational Research Association, Vancouver, Canada.

Cabrera, L. (2019). Teacher preconceptions of computational thinking: A systematic literature review. *Journal of Technology and Teacher Education*, *27*(3), 305–333.

Clarke-Midura, J., Silvis, D., Shumway, J. F., Lee, V. R., & Kozlowski, J. S. (2021). Developing a kindergarten computational thinking assessment using evidence-centred design: The case of algorithmic thinking. *Computer Science Education*, *31*(2), 117–140. doi:10.1080/08993408.2021.1877988

Corradini, I., Lodi, M., & Nardelli, E. (2017). *Conceptions and misconceptions about computational thinking among Italian primary school teachers*. Proceedings of the 2017 ACM Conference on International Computing Education Research - ICER '17, Tacoma, WA, USA. 10.1145/3105726.3106194

Da Cruz Alves, N., Gresse Von Wangenheim, C., & Hauck, J. C. (2019). Approaches to assess computational thinking competences based on code analysis in K-12 education: A systematic mapping study. *Informatics in Education*, *18*(1), 17–39. doi:10.15388/infedu.2019.02

Fraillon, J., Ainley, J., Schulz, W., Friedman, T., & Duckworth, D. (2019). *Preparing for life in a digital world: IEA International Computer and Information Literacy Study 2018 International Report*. International Association for the Evaluation of Educational Achievement. doi:10.1007/978-3-030-38781-5

Gong, D., Yang, H. H., & Cai, J. (2020). Exploring the key influencing factors on college students' computational thinking skills through flipped-classroom instruction. *International Journal of Educational Technology in Higher Education*, *17*(19), 19. doi:10.118641239-020-00196-0

Grover, S. (2018). The 5th 'C' of 21st-century skills? Try computational thinking (not coding). *EdSurge News*. https://edtechbooks.org/-Pz

Grover, S., & Pea, R. (2013). Computational thinking in K–12: A review of the state of the field. *Educational Researcher*, *42*(1), 38–43. doi:10.3102/0013189X12463051

Grover, S., & Pea, R. (2018). Computational thinking: a competency whose time has come. Computer Science Education: Perspectives on teaching and learning in school (pp. 19–37). London. *Bloomsbury Academic., 1033142*. doi:10.1080/08993408.2015

Guggemos, J. (2021). On the predictors of computational thinking and its growth at the high-school level. *Computers & Education, 161*, 104060. doi:10.1016/j.compedu.2020.104060

Hennico, J., Reuter, R. A. P., & Weinberger, M. (2021). Factors Affecting the Teaching of Computational Thinking in Fundamental Schools: A Path Analysis. A paper presented at the *15th Annual Conference of the European Association for Practitioner Research on Improving Learning*. University of Luxenberg. https://hdl.handle.net/10993/48689

Hohlfeld, T. N., Ritzhaupt, A. D., Dawson, K., & Wilson, M. L. (2017). An examination of seven years of technology integration in Florida schools: Through the lens of the levels of digital divide in Schools. *Computers & Education, 113*, 135–161. doi:10.1016/j.compedu.2017.05.017

ISTE & CSTA. (2011). Operational definition of computational thinking for K–12 Education. ISTE. Indiana Government. http://www.iste.org/docs/ctdocuments/computational-thinking-operational-definition flyer.pdf

Jenner, K. (n.d.). *Indiana Academic Standards: Science and Engineering Practices.* Department of Education. https://media.doe.in.gov/news/science-and-engineering-practices.pdf

Kotsopoulos, D., Floyd, L., Khan, S., Namukasa, I. K., Somanath, S., Weber, J., & Yiu, C. (2017). A pedagogical framework for computational thinking. *Digital Experiences in Mathematics Education, 3*(2), 154–171. doi:10.100740751-017-0031-2

Lee, I., Martin, F., Denner, J., Coulter, B., Allan, W., Erickson, J., Malyn-Smith, J., & Werner, L. (2011). Computational thinking for youth in practice. *ACM Inroads, 2*(1), 32–37. doi:10.1145/1929887.1929902

Mcgowan, V. C., Klein, E., & Morrison, D. (2019). *Teaching Tools for Science, Technology, Engineering and Math (STEM) Education.* STEM Teaching Tools. https://stemteachingtools.org/brief/56

Miller, E. C., Severance, S., & Krajcik, J. (2020). Connecting Computational Thinking and Science in a US Elementary Classroom. In J. Anderson & Y. Li (Eds.), *Integrated Approaches to STEM Education: An International Perspective* (pp. 185–204). Springer, Cham. doi:10.1007/978-3-030-52229-2_11

National Research Council. (2010). *Report of a Workshop on the Scope and Nature of Computational Thinking*. Washington, DC: The National Academies Press. https://doi.org/10.17226/12840

Ogegbo, A. A., & Ramnarain, U. (2022). A systematic review of computational thinking in science classrooms. *Studies in Science Education, 58*(2), 203–230. doi:10.1080/03057267.2021.1963580

Peel, A., Sadler, T. D., & Friedrichsen, P. (2022). Algorithmic Explanations: An Unplugged Instructional Approach to Integrate Science and Computational Thinking. *Journal of Science Education and Technology, 31*(4), 428–441. doi:10.100710956-022-09965-0 PMID:35535350

Peters-Burton, E., Rich, P. J., Kitsantas, A., Laclede, L., & Stehle, S. M. (2021). High School Science Teacher Use of Planning Tools to Integrate Computational Thinking. *Journal of Science Teacher Education*. doi:10.1080/1046560X.2021.1970088

Rahayu, T., & Osman, K. (2019). Knowledge level and self-confidence on the computational thinking skills among science teacher candidates. *Jurna Ilm. Pendidik. Fis. Al-Biruni, 8*(1), 117–126. doi:10.24042/jipfalbiruni.v8i1.4450

Rich, P. J., Jones, B., Belikov, O., Yoshikawa, E., & Perkins, M. (2017). Computing and engineering in elementary school: The effect of year-long training on elementary teacher self-efficacy and beliefs about teaching computing and engineering. *International Journal of Computer Science Education in Schools, 1*(1), 1–20. doi:10.21585/ijcses.v1i1.6

Román-González, M., Moreno-León, J., & Robles, G. (2019). Combining assessment tools for a comprehensive evaluation of computational thinking interventions. In S-C, Kong & H. Abelson (Eds.), Computational thinking education (pp. 79-98). Singapore: Springer. doi:10.1007/978-981-13-6528-7_6

Saidin, N. D., Khalid, F., Martin, R., Kuppusamy, Y., & Munusamy, N. A. (2021). Benefits and challenges of applying computational thinking in education. *International Journal of Information and Education Technology (IJIET), 11*(5), 248–254. doi:10.18178/ijiet.2021.11.5.1519

Sands, P., Yadav, A., & Good, J. (2018). Computational thinking in K-12: In-service teacher perceptions of computational thinking. In M. S. Khine (Ed.), *Computational thinking in the STEM disciplines* (pp. 151–164). Springer. doi:10.1007/978-3-319-93566-9_8

Schwab, K. (2016). *The fourth Industrial Revolution*. World Economic Forum. https://luminariaz.files.wordpress.com/2017/11/the-fourth-industrial-revolution-2016-21.pdf

Seufert, S., Guggemos, J., & Sailer, M. (2021). Technology-related knowledge, skills, and attitudes of pre-and in-service teachers: The current situation and emerging trends. *Computers in Human Behavior, 115*, 106552. doi:10.1016/j.chb.2020.106552 PMID:32921901

Stewart, W. H., Baek, Y., Kwid, G., & Taylor, K. (2021). Exploring Factors That Influence Computational Thinking Skills in Elementary Students' Collaborative Robotics. *Journal of Educational Computing Research, 59*(6), 1208–1239. doi:10.1177/0735633121992479

Tang, X., Yin, Y., Lin, Q., Hadad, R., & Zhai, X. (2020). Assessing computational thinking: A systematic review of empirical studies. *Computers & Education, 148*, 103798. doi:10.1016/j.compedu.2019.103798

The Council of State Science Supervisors. (n.d). Vision for Science Education - A Framework for K-12 Science Education: Practices, Crosscutting Concepts, and Core Ideas. COSSS. https://cosss.wildapricot.org/resources/Pictures/Overview/FRPracticesAware1hr.pdf

Tsakeni, M. (2021). Preservice teachers' use of computational thinking to facilitate inquiry-based practical work in multiple-deprived classrooms. *EURASIA Journal of Mathematics, Science and Technology Education, 17*(1). https://doi.org/ doi:10.29333/ejmste/9574

Tucker-Raymond, E., Cassidy, M., & Puttick, G. (2021). Science teachers can teach computational thinking through distributed expertise. *Computers & Education, 173*, 104284. doi:10.1016/j.compedu.2021.104284

Weese, J. L., & Feldhausen, R. (2017). *STEM Outreach: Assessing Computational Thinking and Problem-Solving*. 124th American Society for Engineering Education Annual Conference and Exposition (ASEE 2017), Columbus, OH, USA. https://peer.asee.org/stem-outreach-assessing-computational-thinking-and-problem-solving

Weintrop, D., Beheshti, E., Horn, M., Orton, K., Jona, K., Trouille, L., & Wilensky, U. (2016). Defining computational thinking for mathematics and science classrooms. *Journal of Science Education and Technology, 25*(1), 127–147. doi:10.100710956-015-9581-5

Weintrop, D., Rutstein, D. W., Bienkowski, M., & McGee, S. (2021). Assessing computational thinking: An overview of the field. *Computer Science Education, 31*(2), 113–116. doi:10.1080/08993408.2021.1918380

Wing, J. (2006). Computational thinking. *Communications of the ACM*, *49*(3), 33–35. doi:10.1145/1118178.1118215

Yadav, A., Krist, C., Good, J., & Caeli, E. N. (2018). Computational thinking in elementary classrooms: Measuring teacher understanding of computational ideas for teaching science. *Computer Science Education*, *28*(4), 371–400. doi:10.1080/08993408.2018.1560550

Yin, Y., Hadad, R., Tang, X., & Lin, Q. (2020). Improving and assessing computational thinking in maker activities: The integration with physics and engineering learning. *Journal of Science Education and Technology*, *29*(2), 189–214. doi:10.100710956-019-09794-8

Zhong, B., Wang, Q., Chen, J., & Li, Y. (2016). An exploration of three-dimensional integrated assessment for computational thinking. *Journal of Educational Computing Research*, *53*(4), 562–590. doi:10.1177/0735633115608444

Chapter 5
The Affordances of Virtual Learning Environments in Remote Chemistry Learning

Mafor Penn
University of Johannesburg, South Africa

Umesh Ramnarain
University of Johannesburg, South Africa

ABSTRACT

Today, virtual learning environments (VLEs) are increasingly valuable for mediating teaching and learning in higher education. This chapter reports the affordances of VLEs in remote chemistry learning. The effects of VLEs versus remote lectures on students' achievement scores in chemistry content tests were compared, and the affordances of VLEs for 21st-century chemistry learning were investigated. This research aimed to investigate the impact of VLEs when learning chemistry remotely and answer some questions about their affordances in the fourth industrial revolution (4IR). Fifty pre-service science teachers (also referred to as students) from a South African university participated in the studies. A mixed methods approach with a quasi-experimental design was preferred. Pre-and post-tests were used to evaluate achievement scores for control and experimental groups, while focus group interviews were used for data gathering in reflection. Results showed a positive shift in achievement scores of the experimental group compared to the control group.

DOI: 10.4018/978-1-6684-6932-3.ch005

INTRODUCTION

At the onset of the COVID-19 pandemic, several students globally were compelled to learn in diverse virtual learning environments (VLEs) due to the restrictions imposed by a series of lockdowns. VLEs are learning spaces, usually online or offline, embedded with simulations, demonstrations and illustrations of scientific processes and systems (Limniou & Smith, 2010). What makes VLEs unique is the addition of learning activities that scaffold a learner's use and interactions with the embedded content (Chamberlain et al., 2014). The experiences of the global response to the COVID-19 pandemic enforced the consideration of new ways of teaching and learning in several educational settings (Asim et al., 2020). While the world as we know it will never revert to pre-COVID-19 conditions, VLEs have become more attractive for researchers and educators alike (Alves et al., 2017). As the world increasingly transforms into a digital village, using available technological tools in science teaching and learning is inevitable for teachers and students alike. This is because 21st-century learners tend to rely on technology to attain diverse learning goals (Alves et al., 2017; Reisoğlu, et al., 2017).

In the case of chemistry education, where laboratory-based learning is a must-have and abstract concepts are usually difficult to comprehend from just engaging with text, diagrams and equations, it becomes questionable how VLEs can help students learn the subject in remote settings. Learning difficulties in chemistry typically revolve around students' poor visualisation of concepts like atoms, electrons, reactivity, molecules, and other micro-scientific phenomena (Faour & Ayoubi, 2018). This chapter firstly breaks down a series of sequential findings from first comparing VLE-based chemistry learning experienced by an experimental group of pre-service teachers (PST) against a control group of PSTs that engaged in lectures only. Secondly, the chapter looks at the same group of participant PSTs' perceptions and reflections on the affordances VLEs for 21st-century science learning. Some future directions are also suggested based on this study.

BACKGROUND

The working definition of VLES in the current chapter is as follows: A virtual learning environment (VLE) is an educational Web-based platform for the digital aspects of a course or module, usually situated in an institutional platform. VLEs present resources, activities and interactions within the module structure and provide for the different stages of assessment. VLEs also usually report on participation; and have some level of integration with other institutional systems (Britain & Oleg, 1999).

Virtual learning environments (VLEs) are different from virtual classrooms, which replicate physical classrooms to enable people to learn from a distance and connect with content delivery. VLEs are intended to harness technological tools that enhance teaching and learning experiences in virtual spaces (Boulton et al., 2018). In a typical VLE, the instructor tends to include different virtual artefacts, including virtual laboratories, videos, simulations, and assessments for students to attempt in the learning process.

The ultimate goal of adopting such a learning environment is to enhance active engagement with the content, interactivity and in science education inquiry-based learning opportunities for students when they learn remotely. In a nutshell, a VLE helps instructors to create and facilitate technology-enhanced learning for students while developing learning autonomy (Alves et al., 2017; Britain & Liber, 1999). In institutions of higher learning, VLEs are usually integrated within broader institutional learning management systems (LMS).

VLEs have contributed immensely to extending face-to-face learning, especially in science education, for the last two decades. Some researchers, including Alhalabi (2016), De Jong (2017), and Kanimozhi. (2018) argue that VLEs and the diverse artefacts embedded in them help students improve engagement, visualisation and interactivity with scientific phenomena. These engagements in VLEs eventually enhance conceptual understandings as well as build procedural skills and the mastery of processes (Al-Nuaimy et al., 2021; Alves et al., 2017).

Also, once a VLE has been well-designed and scaffolded, students will inherently develop autonomous and self-directed learning skills (Alves et al., 2017).

Another benefit of creating and using VLEs in chemistry instruction and learning is that they provide a safe virtual space for experimentation (Penn & Ramnarain, 2019). Within VLEs, students can use technological tools to conduct dangerous experiments. In a VLE, these kinds of experiments can be repeated, mastered and understood before one engages with them in a face-to-face laboratory.

Learning remotely with VLEs also adjusts the teacher-to-student ratio, where tasks can be simultaneously carried out in virtual laboratories without dealing with overcrowding issues and lack of instructor attention to the process (Zeichner & Zilka, 2016).

In VLEs, traditional ideas about teaching and learning upheld in the teacher-centred model are relegated. The focus in the VLE then becomes knowledge sharing rather than knowledge hoarding or transmission. VLEs provide a platform for students to look at the future in which knowing how to learn autonomously will be a fundamental skill, given the pace of innovation and the changing structure of workplaces and communities.

Despite these accrued benefits reported on VLE-enhanced learning and instruction, a comparative study of achievement scores from VLE-enhanced against traditional lectures is crucial in PST chemistry education, especially in the South African context where comparative studies of this nature are scarce. Hence, this study investigates the possible effects of using VLEs versus traditional instruction on PSTs' achievement in chemistry content tests at a South African university. The second phase of the study looks at students' perceptions and reflections on the affordances of VLEs in chemistry learning. The following research questions drive the sequence of investigations:

- What are the effects of remote VLE-enhanced versus traditional instruction on PSTs' achievement in chemistry?
- How do PSTs perceive the use of VLEs in remote chemistry learning?

Aims and Objectives

The main aim of the study was to compare the affordances of VLEs to traditional instructional models like lectures by assessing PSTs' achievement scores in control and experimental groups. Then, further explicate the affordances of VLEs by looking at PSTs' perceptions and reflections on their experiences.

Based on these aims, the following objectives were set:

- To assess and compare students' achievements in chemistry content test pre and post-VLE/traditional chemistry instruction
- To investigate PSTs' perceptions of the affordances of VLEs in chemistry learning.

Hypotheses Stated

The Null Hypothesis (H_0): There is no difference in post-test achievement scores for control and experimental groups of pre-service chemistry teachers.

The Alternative Hypothesis (H_a): There is a difference in post-test achievement scores for control and c experimental groups of pre-service chemistry teachers.

The rationale for such a chapter is to showcase the affordances of VLE in remote settings and their role as a complementary tool for chemistry and science learning.

Learning Chemistry Remotely

Learning chemistry has always been associated with representational and visualisation difficulties. Most times than not, concepts in chemistry only become explicit by engaging in some form of experimentation in order to make sense of concepts related to reactions, the flow of electrons and other chemico-physical properties associated with the state of matter. In some literature, students have indicated that the complexities of the subject are such that without a platform to visualise what actually happens in certain reactions, making mental representations of concepts becomes relatively daunting (Adesoji et al., 2017; Faour & Ayoubi, 2018; Tümay, 2016). In essence, chemistry is a practical subject. For it to appeal to the human senses, there should be multiple representations of concepts (Adesoji et al., 2017).

At the start of the global pandemic in the year 2020, several lockdown regulations prohibited students from laboratory-based instruction in tertiary chemistry modules. As a result, instructors globally had to be proactive in designing suitable VLEs where similar laboratory-based learning could take place. In this current research, VLEs were embedded within the broader LMS whereby selected applications were used to facilitate virtual laboratory-based chemistry learning for PSTs to assess VLE impact and provide the much-needed leverage for practical chemistry. Some of the fundamental questions that kept the researchers thinking included:

Which pedagogical strategies can be used to facilitate learning in VLEs? How can concepts be scaffolded in VLEs? How can assessments be facilitated and progress measured in VLEs? Overarchingly what are students' perceptions of the affordances of VLEs? All of these questions were important because it is only when students have a positive perception of such learning environments that learning can be enhanced using them.

Theoretical Background

In answering questions related to the design of a functional VLE for chemistry teaching and learning, one needs to examine different pedagogical strategies and activities that could be embedded in the VLE teaching and learning theories to ensure their effectiveness. In principle, a VLE could become a space for an interactive educational engagement where students can explore problem-solving, investigations and creativity in their learning of concepts (Reisoğlu, et al. 2017). These attributes are associated with constructivism, a theory that favours learner-centred knowledge creation (Schunk, 2012). Proponents of constructivism hold that learning is an active process in which the learner participates in knowledge construction by making linkages to their prior knowledge and cultural experiences (Schunk, 2012; Von Glasersfeld, 1989). According to famous constructivist theorists like Vygotsky (1978), it is critical to scaffold learning while still giving the learner an opportunity to navigate from novice to expert thinking across the zone of proximal development (ZPD). VLEs, by the very nature of their content, support learning in constructivist manners and the interactions that come with sharing knowledge in a boundless learning environment. The more radical form of constructivism which borrows from Piaget's (1985) theories, is rooted in explicating diverse pragmatic approaches to explaining reality. This form of constructivism works well for integrating VLEs in teaching and learning as one is given the opportunity to use what works to make meaning of concepts and ideas, especially in science (Walshe, 2020). Based on the aforementioned theoretical underpinnings, strategies like the use of simulations, illustrations, examples, analogies, videos, explanations and virtual demonstrations could be key in engaging students and improving their ability to transform abstract concepts into more concrete ideas as they learn in chemistry.

Another dimension of VLEs is the idea of having different kinds of representations with audio, visual, text, kinesthetic and other forms of animated instructions like simulations that could facilitate an interactive learning experience within the VLE. These attributes are also strongly associated with Mayer's cognitive theory of multimedia learning (CTMML), a theory which holds that learning will be better attained when multiple representations are used (Mayer, 2011). A good VLE provides a combination of representations with pictures, shapes, text and scenes mixed with sound and movement, which will enable one to learn in ways that are best appealing to them. The fundamental dimensions of the CTMML are adequate in supporting the dynamic nature of VLEs and how learning is fostered within these learning spaces.

METHODOLOGY

The first phase of the study was entirely quantitative. It followed a quasi-experimental design whereby the researchers worked with a specific group of fifty (n=50) PSTs randomly assigned to control and experimental groups. Participant PSTs included 35 males and 15 females in the third year of a Bachelor of Education (B.Ed) programme. The age group ranged from 20-23, and participant PSTs had a sound knowledge of using the institutional LMS Blackboard. Pseudonyms were given to the participants (S1-S50). S1-S25 constituted the control group, while S26-S50 made up the experimental group.

Nature of Remote Intervention With VLEs

In commencing the remote learning interventions for PSTs in the experimental group, an experimental content folder hosting the designed VLE was created within the existing LMS Blackboard in a Physical Sciences course. The created content folder was selectively assigned only to the experimental group through an adaptive release process, and those of the control group had no access to it.

The researchers identify useful chemistry applications to embed in the VLE against the topics and learning outcomes from the module. Specifically, an open-source chemistry application with several experiments associated with chemical reactions, videos and PhET simulations all constituted part of the experimental VLE. Open-ended and multiple-choice assessments, structured to meet the specific learning outcomes, were also designed and embedded in the VLE at the end of each learning milestone. The VLE engagements included group work, discussion boards, making blog posts and group meeting rooms on Blackboard.

The interventions were targeted at different aspects of chemical reactions over three weeks. Concepts included: types of chemical reactions, electronegativity/electropositivity and reactivity. The control group attended traditional remote lectures where slides were presented and concepts explained with no additional virtual laboratories or simulations. The experimental group engaged with the embedded VLE folder with the same slides as the control group and minimal facilitation from the module lecturer.

Using an inquiry-based approach to the instruction, PSTs could select the reactants in a combustion, oxidation or reduction reaction to investigate and report on. After milestone investigations and activities are completed in the VLE folder, a PST's progress is automatically recorded, and badges are awarded in the case of the top score.

At the end of the three-week cycle of interventions, post-content tests were completed, and the PSTs in the control group were flipped and given access to the experimental VLE folder. The same facilitation was provided to the control group to ensure equitable learning opportunities for all participant PSTs.

Data Collection and Analysis

Data was collected by means of administering a chemistry content-based pre-test to all PSTs prior to engagement with the actual content by VLE/traditional lecture instruction. The test completion time was 90 minutes. All collected test data were analysed by statistical means using SPSS. For ethical reasons, those who participated in the control group were at a later post-experimental stage exposed to the same VLE in order to ensure equitable learning experiences for all PSTs and also to include all PSTs in the focus group interviews in the second study.

In the second phase of the study, participating PSTs were grouped into ten (five PSTs per group) sub-groups to ease participation in focus group interviews. The main aim of the focus group interviews was to establish PSTs' perceptions of the affordances of VLEs. Five main interview questions were posed to all participants, including:

- What are your overall experiences with VLE-enhanced chemistry learning?
- What were the highlighted benefits you experienced?
- What do you think might have been the reason for the benefits experiences?

- What were the highlighted challenges that you experienced? Provide explanations.
- In your own words, reflect on the affordances (possibilities) of VLEs in chemistry learning.

The interviews were audio recorded, transcribed into text, inductively coded and analysed using thematic content analysis. As the analysis proceeded, combined codes became categories and combined categories became themes (Merriam & Tisdell, 2016). From the generated themes, the researchers presented five themes on PSTs' experiences of using VLEs in remote chemistry learning scenarios.

FINDINGS

Phase One

Results from the quantitative phase of this study are presented here. Two groups of PSTs, control (n=S1-S25) and experimental (n=S26-S50), wrote pre-test on concepts related to chemical reactions without prior treatments to content knowledge. A three-week period of lecture and VLE-enhanced learning interventions were completed, and the post-test content test was completed. Table 1 presents the descriptive statistics and shows the average mean scores (*M*) and standard deviations (*SD*) of the two groups' pre-and post-test achievement scores.

Table 1. Pre and post-test mean scores for groups

	Pre-Service teacher groups	N	Mean	Std. Deviation	Std. Error Mean
Pre-tests	Control group	25	59.44	8.84	1.77
	Experimental group	25	59.84	9.40	1.88
Post-test	Control group	25	67.92	8.58	1.72
	Experimental group	25	77.24	7.46	1.49

From Table 1 pre-test mean scores of the two groups of PSTs were almost the same, with the control group having a mean score of (M= 59.44%, S.D= 8.84) while the experimental group had a pre-test mean score of (M=59.84%, S.D = 9.40). The mean score difference between the groups was computed to be 0.4% (59.84-59.44 = 0.4). Upon establishing that the pre-test data were normally distributed, a further paired sample t-test test confirmed that the differences in the pre-test mean scores for the two groups were not statistically significant *($t(48)= -1.16, p> 0.05=.88$)* at 95% confidence interval. This result indicates that before the VLE-enhanced instructions, PSTs of these groups had the same ability and achievement levels in chemistry.

On the other hand, the Post-test mean score for the experimental group was higher at (M = 77.24%, SD = 7.46) than the post-test mean score of the control group at (M= 67.92%, SD = 8.58). The observed mean difference between the two groups was computed to be about 9.32% and therefore had to be verified for statistical significance. The post-test data was also established to be normally distributed.

Comparative Analysis of Groups Using Independent T-Tests

Table 2 displays the findings from paired samples t-test conducted to establish how statistically significant the difference in the post-test mean scores for the experimental and control groups were.

Table 2. Independent samples t-test results for differences in group pre and post-test scores

Groups	N	df	Std. E Mean	t	p
Pre-Control	25	48	2.58	-1.16	p> .05= .88
Pre-Experimental	25				
Post-Control	52		2.27	-4.10	P < .001
Post-experimental	52	48			

As seen in Table 2, the mean post-test scores for the experimental group of PSTs were significantly higher ($t(48) =-4.10$, $P<.01$) than the post-test score of the control group at a 95% confidence interval. Pre-test means showed no statistically significant differences at a 95% confidence interval with ($t(48)= -1.16$, $p=.88$).

These Post-test results suggested a significant improvement in students' achievement in chemistry content tests after the intervention with VLEs compared to remote lectures. Embedded VLE in the course clearly had an effect on the level of engagement that PSTs had with the content, and as expected, this affected their achievement scores. Hence the alternative hypothesis (H_a) is accepted, while the null hypothesis (H_0) is rejected for this study. It may also be possible that other factors might have affected the post-test scores of experimental group participants. However, the effect size calculated within the software at d=1.12 for the post-test data against the effect size d= 0.04 for the pre-test data indicated that, to a larger extent, the positive shifts for both control and experimental groups emanated from the learning interventions.

Phase Two

The results of the qualitative phase of the study are presented here. In the follow-up semi-structured focus group interview, PSTs' experiences were reflected on to obtain their perceptions of the affordances of VLEs used in chemistry learning. From the thematic content analysis of interview data, five main themes which describe the holistic affordances of VLEs were arrived at.

Well Scaffolded Learning Activities in VLEs Promote Autonomous Learning

PSTs indicated that the different learning activities, videos, simulations, virtual laboratories, animations and the associated assessments task therein enhanced their abilities to pace their learning and learn independently of direct instructions from the lecturer to a large extent. For example, S40 indicated, *"I found the VLE to provide a clear path to follow from the introduction of a topic till the end. With milestone assessments at different levels, it was easier for me to self-teach and self-assess, which is empowering."* S21 from the control group, who only engaged with the VLE after the post-test, indicated that *"I feel that I needed this kind of interaction with the VLE. Comparatively, I found that there was more in the VLE folder, which helped me to learn things on my own and even faster. The simulations in the VLE took my learning to the next level, and I had an eye-opening experience."*

Overall from the focused group interviews, PSTs indicated that well-scaffolded learning activities in VLEs and the embedded systemic assessments and feedback in the applications included were effective for self-learning, self-assessment and trial and error in pre-lab preparations. The learning environment was easy to navigate, and one could learn effectively with minimal external instruction from a tutor.

Investigations in Chemistry Are Feasible in VLEs

PSTs indicated that facilitated virtual chemistry laboratories and some of the PhET simulations that were embedded in the VLE provided the opportunity for them to engage with experimental chemistry as well as investigate concepts. These VLE tools enhanced their visualisation of abstract chemical reactions and the interactions between sub-atomic particles like electrons and charges. All ten groups of students indicated that experimenting with VLEs was crucial for inquiry-based learning and the visualisation of chemical reactions in remote settings. For instance, S31 said, *"I can't imagine what would happen with us learning the different types of reaction without engaging with some of the virtual labs. This VLE experience really helped me to see exactly what goes on with valence shell electrons in certain reactions."* S29 agreed and said, *"I agree with S31. In this group, we were able to investigate the reactions between different kinds of reactants that were not even in the scope of the learning tasks, and we were shocked at some of our observations (laughing interlude)."*

This theme was clearly associated with one of the fundamental affordances of VLEs in that they are able to host virtual laboratories, which can augment theoretical learning experiences for chemistry students learning remotely.

VLEs Aid the Development of Higher-Order Thinking Skills (HOTS)

Looking at the VLE interphases in this study, PSTs had embedded discussion boards for chats. They could create blogs, engage in group work, and direct conversations with each other using *Collaborate,* a platform on the host LMS Blackboard. All of these avenues contributed to participants' learning experiences such that skills like collaboration, communication, creativity and critical thinking could be developed. One participant, S28, indicated that *"I find the VLE to be a flexible learning environment as it gave me room to be creative and try my own thoughts out. I was able to design a chemist blog on the applications of reduction reactions in real life. I am not sure I could have thought of this had it not been for the VLE."* From the control group, participant S14 said, *"I feel that discussion boards on the VLEs enhanced my communication and collaboration skills with my peers. Whenever I encountered difficulties, I would pose a question on the discussion boards, and the responses from group members will be interactive and stimulating of one's thought processes."*

All of these aspects made VLEs a suitable resource for developing some higher other thinking skills (HOTS) in the participant PSTs.These skills are crucial for learning and work in the fourth industrial revolution and should be directly or indirectly nurtured in young people.

VLEs Support Remote Learning and Access to Global Learning Resources

For participant PSTs in this study, VLEs made learning convenient in remote settings. With restrictions on access to chemistry laboratories during the remote learning periods, PSTs could still conduct certain experiments virtually and even have access to other supporting learning materials developed by institutions all over the world.

This, in essence, showed that VLEs provided PSTs with possibilities beyond their geographical locations. Some PSTs indicated embedded virtual laboratories in the VLE were more convenient to use than the face-to-face experiences in the physical laboratory in their institution. For example, S35 said, *"I found the VLE to really enhance my love for experimentation than attending labs at varsity. In the physical lab, we are always crowded, and sometimes reagents are not even enough for all the groups. However, with these virtual labs, I feel like an independent scientist being able to do my work seamlessly"*. In supporting the access to global learning resources, S49 indicated, *"The good thing with learning virtually is that I had access to a wide variety of content from different institutions available to me through the click of a mouse. I really feel that VLEs are the future of education in the fourth industrial revolution."* These findings suggest that VLEs have the ability to expose students to many possibilities beyond the geographical location they are based in. The results also suggest that there is convenience and safety experienced within VLEs when experimenting with chemical reactions.

VLEs are a Complementary Tool to Traditional Instructions

Participant PSTs in this study indicated that in chemistry learning, VLEs could be an excellent complementary tool for traditional face-to-face learning experiences as it provides the opportunity to master procedural skills for experimentation in chemistry as part of pre-laboratory preparation. They also indicated that the VLE provides the opportunity for online learning resources, which could be added to what is obtained through face-to-face engagements. VLEs provide room for more students to learn concepts at once and help the instructor to manage the teacher-student ratio effectively. For example, S15 from the control group indicated that *"If I had a VLE-based laboratory session before I went to the lab every time, I feel that I will do better in all the practical tasks. The VLE gives me the opportunity for me to try many things and even repeat them for mastery."*

S33 also indicated that *"VLEs help with some of the problems of overcrowding that we always experience at the laboratory and reduce the number of laboratory slots per topic since everybody has access to virtual resources at the same time. This, to me, means that time is saved, and the lecturer's load and the need to facilitate too many sessions is reduced"*.

Some Barriers to Using VLEs in Chemistry Learning

As with all types of technology-enhanced learning, there are several challenges associated with using VLEs and PSTs were asked to report back on the challenges they experienced.

PSTs indicated that connectivity issues were experienced on some days when engaging with VLEs. Some of the virtual laboratories and simulations embedded in the VLEs had content misrepresentations, which could lead to misconceptions if they had not been well-scaffolded in the VLEs. Sometimes it was the technical issues that were associated with certain applications within the VLE that marred the experience. One other contextual challenge was the power cuts that would make virtual content inaccessible for some hours of the day. Some quotations from the participant PSTs supported the points raised in terms of challenges. For example, S28 indicated that *"learning virtually was challenging in my area on some days as the internet connectivity would fluctuate, increasing the lag time in opening certain applications"* On the aspect of misrepresentations, S45 said, *"Some of the experiments within the virtual chemistry lab were not adequately represented as I struggled to see the gas generated as parts of the products of the reactions.*". These findings suggest that there are several challenges that are associated with VLE-enhanced learning. Some of them could be technical, contextual or content-related. Hence instructors must be adequately prepared to address these challenges.

DISCUSSIONS AND CONCLUSION

The findings of this study based on the obtained results suggest that VLEs can enhance students' understandings of chemistry concepts due to their interactive nature. PTSs indicated that remote chemistry learning could only improve with the additional VLE resources. PSTs noted that embedded learning tools, such as simulations, videos, illustrations, chats, discussion boards and blogs, enhanced understanding of science concepts through interactivity and peer collaborations. Students ultimately reflected that remote learning is convenient learning and reduces the need for face-to-face interactions with tutors or lecturers.

These findings are congruent with findings of other studies like Zeichner and Zilka (2016), who reported that once the usability issues associated with the initial use of VLEs have been dissipated, students enjoyed positive learning experiences within virtual learning spaces. Zacharia and Olympiou (2011) also found that for physics learning, conceptual understandings are enhanced when students combine traditional laboratory learning with interactive virtual learning, as is the case when learning with VLEs. Enhanced student achievements are also reported in VLEs as a result of some of the characteristics like interactivity and convenience, which are inherent to them (Alves et al., 2017). A study by Kizilcec et al. (2017) also suggests that VLEs provide access to a wide array of educational resources, such as animations, videos and simulations which contribute to learning positively with multiple representations.

Contrary to these findings, some studies, like Wright et al. (2014), reported that students who received traditional classroom chemistry instruction outperformed those who received VLE-based. These findings were explained by the lack of hands-on laboratory engagement and face-to-face interactions. Other researchers suggest that learning in VLEs is dependent on the level of students' maturity and experience with using such learning environments. For example, Park and Kim (2020), as well as Ruthotto et al. (2020), report that more experienced students, like the participants in this study, will tend to perform better when learning in Vles than freshers as they understand the settings better. They argue that such older students are better at communication in these spaces, they are motivated and can plan better for self-directed learning (Kauffman, 2015; Ruthotto et al., 2020)

Referring to the affordances of learning science with VLEs, especially in experimental science modules like chemistry. PSTs highlight that VLEs are essential for developing autonomous learning skills, virtual experimentation and higher-order thinking skills. From PSTs' experiences, VLEs are complementary tools to traditional chemistry instruction and can adequately support access to global learning resources in remote learning scenarios. This is congruent with contemporary research to a great extent. For example, researchers like Shuja (2019) and Zawacki-richter et al. (2019) report on the flexibility, autonomy and convenience of learning in virtual spaces in association with the futuristic skills that come with online learning.

Reporting on some of the challenges, PSTs mentioned content representation issues, technical issues with certain virtual laboratories and simulations and contextual issues related to power failure. Dean and Lima (2017) also report similar setbacks in adopting and using VLEs. To overcome these challenges, instructors need time and support to plan and create pedagogically worthy VLEs adequately. Other challenges were related to the lack of interpersonal and social factors like peer relations and emotional connections, as also reported in studies like Burde et al. (2017). The skills to do this must be learned, and one must be adaptable to changing educational technologies.

Based on the analysis of these findings, it is clear from our perspective that the integration of VLEs in chemistry in remote or online settings is an educational solution that must be well-planned, scaffolded and rich in multimedia resources. With the extensive availability of virtual laboratory resources, simulations and models, successful VLEs can be created where students will benefit immensely and well above the use of traditional resources only.

These findings have implications for adapting VLEs as complementary instructional media to in-person chemistry learning. Nevertheless, students must engage in traditional practical work in chemistry laboratory settings in order to develop authentic science laboratory process skills in preparation for the world of work.

LIMITATIONS AND IMPLICATIONS

The study was limited by participant numbers and the focus on one higher education institution in a particular context, so the results are not generalisable. It is also relevant to stress that there may be other factors that affect learning in VLEs that may not be within the researchers' control. The results, however, open windows of consideration for preparing PSTs for their future practice. The achievement score shifts indicate that there is value in well-organised VLEs for improving performance in chemistry. Therefore subjecting prospective teachers to different VLEs accustoms them to other ways of teaching and learning in the fourth industrial revolution in preparation for their practice.

Based on these findings of the current study, it is recommended that larger-scale studies be employed to investigate further the affordances of VLEs in Chemistry teacher education, including all the different factors that affect how students learn in VLEs.For science teacher educators, remote learning ideas must be nurtured in PST education for the 21st century in compliance with the changing educational paradigm globally. Lessons learnt from the COVID-19 pandemic suggest that learning in s safe physical space may not always be guaranteed. Hence the adoption and utilisation of virtual learning spaces are imperative.

REFERENCES

Adesoji, F. A., Omilani, N. A., & Dada, S. O. (2017). A comparison of perceived and actual; Students' learning difficulties in physical chemistry. *International Journal of Brain and Cognitive Sciences*, *6*(1), 1–8.

Al-Nuaimy, E. A., Twaij, A. H., & Mahmood, H. A. (2021). Importance of Distance Education and its Rise during the Period COVID-19. *International Journal of Educational Research*, *12*(2), 29–34.

Alhalabi, W. (2016). Virtual reality systems enhance students' achievements in engineering education. *Behaviour & Information Technology*, *35*(11), 919–925. do i:10.1080/0144929X.2016.1212931

Alves, P., Miranda, L., & Morais, C. (2017). The influence of virtual learning environments in students' performance. *Universal Journal of Educational Research*, *5*(3), 517–527. doi:10.13189/ujer.2017.050325

Asim, S., Ponners, P. J., Bartlett, C., Parker, A., & Star, R. (2020). Differentiating Instruction: For Middle School Students in Virtual Learning Environments. *Delta Kappa Gamma Bulletin*, *86*(3), 19–31.

Blaschke, L. M. (2012). Heutagogy and lifelong learning: A review of heutagogical practice and self-determined learning. *International Review of Research in Open and Distance Learning*, *13*(1), 56–71. https://www.irrodl.org/index.php/irrodl/article/view/1076/2113. doi:10.19173/irrodl.v13i1.1076

Boulton, C. A., Kent, C., & Williams, H. T. (2018). Virtual learning environment engagement and learning outcomes at a `bricks-and-mortar' university. *Computers & Education, 126*, 129–142. doi:10.1016/j.compedu.2018.06.031

Britain, S., & Liber, O. (1999). *A Framework for Pedagogical Evaluation of Virtual Learning Environments*. Department of Education.

Burde, D., Kapit, A., Wahl, R. L., Guven, O., & Skarpeteig, M. I. (2017). Education in emergencies: A review of theory and research. *Review of Educational Research, 87*(3), 619–658. doi:10.3102/0034654316671594

Chamberlain, J. M., Lancaster, K., Parson, R., & Perkins, K. K. (2014). How guidance affects student engagement with an interactive simulation. *Chemistry Education Research and Practice*, *15*(4), 628–638. doi:10.1039/C4RP00009A

De Jong, T. (2017). Instruction based on computer simulations and virtual laboratories. In R. E. Mayer & P. A. Alexander (Eds.), *Handbook of research on learning and instruction* (2nd ed., pp. 502–521). Routledge.

Dean, A., & Lima, A. (2017). Student Experience of E-Learning Tools in HE: An Integrated Learning Framework. *European Journal of Social Sciences Education and Research*, *11*(2), 39–51. doi:10.26417/ejser.v11i2.p39-51

Faour, M. A., & Ayoubi, Z. (2018). The effect of using virtual laboratory on grade 10 students' conceptual understanding and their attitudes towards physics. [JESEH]. *Journal of Education in Science, Environment and Health*, *4*(1), 54–68. doi:10.21891/jeseh.387482

Kanimozhi. (2018). Virtual Learning Environment. *International Journal of Advanced Research in Computer Science, 9*(1), 11–14.

Kauffman, H. (2015). A review of predictive factors of student success in and satisfaction with online learning. *Research in Learning Technology*. doi:. v23. 26507. doi:10.3402/rlt

Kizilcec, R. F., Piech, C., & Schneider, E. (2017). Deconstructing disengagement: Analyzing learner subpopulations in massive open online courses. In Proceedings of the Fourth (2017) *ACM Conference on Learning @ Scale* (pp. 303-312). ACM.

Limniou, M., & Smith, M. (2010). Teachers' and students' perspectives on teaching and learning through virtual learning environments. *European Journal of Engineering Education, 35*(6), 645–653. doi:10.1080/03043797.2010.505279

Mayer, R. E. (2011). Instruction based on visualisations. In *Handbook of research on learning and instruction* (pp. 441–459). Routledge. doi:10.4324/9780203839089-30

Merriam, S., & Tisdell, E. (2016). Qualitative research: A guide to design and implementation (4th ed.) John Wiley & Sons, Inc.

Park, C., & Kim, D. G. (2020). Exploring the roles of social presence and gender difference in online learning. *Decision Sciences Journal of Innovative Education, 18*(2), 291–312. doi:10.1111/dsji.12207

Piaget, J. (1985) *The equilibration of cognitive structures: The central problem of intellectual development.* University of Chicago Press (original work published 1975).

Penn, M., & Umesh, R. (2019). The use of virtual learning environments and achievement in physics content tests. In *Proceedings of the International Conference on Education and New Development* (pp. 493-497). ACM.

Reisoğlu, I., Topu, B., Yılmaz, R., Karakuş Yılmaz, T., & Göktaş, Y. (2017). 3D virtual learning environments in education: A meta-review. *Asia Pacific Education Review, 18*(1), 81–100. doi:10.100712564-016-9467-0

Ruthotto, I., Kreth, Q., Stevens, J., Trively, C., & Melkers, J. (2020). Lurking and participation in the virtual classroom: The effects of gender, race, and age among graduate students in computer science. *Computers & Education, 151*, 103854. doi:10.1016/j.compedu.2020.103854

Schunk, D. H. (2012). *Learning theories an educational perspective* (6th ed.). Pearson.

Shuja, A. (2019). Effect of E-learning on Students' Academic Performance Mediated by Facilitation Discourse and Flexibility. *Knowledge Management & E-Learning, 11*(2), 158–200.

Tümay, H. (2016). Reconsidering learning difficulties and misconceptions in chemistry: Emergence in chemistry and its implications for chemical education. *Chemistry Education Research and Practice, 17*(2), 229–245. doi:10.1039/C6RP00008H

Von Glasersfeld, E. (1989). Cognition, construction of knowledge, and teaching. *Synthese, 80*(1), 121–140. doi:10.1007/BF00869951

Vygotsky, L. S., & Cole, M. (1978). *Mind in society: Development of higher psychological processes*. Harvard university press.

Walshe, G. (2020). Radical Constructivism—von Glasersfeld. In B. Akpan & T. J. Kennedy (Eds.), *Science Education in Theory and Practice. Springer Texts in Education*. Springer. doi:10.1007/978-3-030-43620-9_24

Wright, T., Barnett, M., Hodge, B., & O'Connell, R. (2014). Investigating the effectiveness of online versus face-to-face chemistry instruction using virtual laboratory simulations. *Journal of Science Education and Technology*, *23*(6), 862–870.

Zacharia, Z. C., & Olympiou, G. (2011). Physical versus virtual manipulative experimentation in physics learning. *Learning and Instruction*, *21*(3), 317–331. doi:10.1016/j.learninstruc.2010.03.001

Zawacki-richter, O., Marín, V. I., Bond, M., & Gouverneur, F. (2019). Systematic Review of Research on Artificial Intelligence Applications in Higher Education –Where Are the Educators? *International Journal of Educational Technology in Higher Education*, *16*(39), 1–27. doi:10.118641239-019-0171-0

Zeichner, O., & Zilka, G. (2016). Feelings of challenge and treat among pre-service teachers studying in different learning environments-virtual vs. blended courses. *Journal of Educational Technology*, *13*(1), 7–19.

Chapter 6
Towards Automatic Analysis of Science Classroom Talk:
Focus on Teacher Questions

Sami Lehesvuori
https://orcid.org/0000-0003-3889-5279
University of Jyväskylä, Finland

Danner Schlotterbeck
University of Chile, Chile

Abelino Jiménez
University of Chile, Chile

Daniela Caballero
McMaster University, Canada

Roberto Araya
University of Chile, Chile

Raija Hämäläinen
University of Jyväskylä, Finland

ABSTRACT

In this study, classroom interaction is addressed by using automatic speech recognition (ASR) for automatic analysis of teacher talk. Educational research on classroom interactions and talk is mainly based on manual analysis, which, although quite accurate, is time consuming and expensive. As there are already many ways to harness the power of automatic approaches to analyse the content of talk, the next step is the automatic detection of different types of talk. In this study, the latter includes the automatic detection of teacher questions and their differentiation into closed and open types. The data consists of ASR text outputs of 25 physics lessons on the same topic, each 90 minutes long, from 25 different science teachers. The interplay between human- and machine-led approaches resulted in promising steps taken towards automatization. Future possibilities for automatic analysis of classroom talk and the implications for teaching and learning are discussed.

DOI: 10.4018/978-1-6684-6932-3.ch006

INTRODUCTION

This chapter seeks to answer the call to reduce the complexity and laboriousness of classroom interaction analysis by modernising and complementing human analysis with novel computational methods. The role instruction and classroom interactions play in the learning process has been addressed in quantitative studies (Fischer et al., 2014; Howe et al., 2019; Müller et al., 2016), but the nature of this connection has not yet been sufficiently addressed due to the complexity and temporal nature of classroom interactions and the laborious methods involved in their analysis. As Hennessy et al. (2020) noted, the time is right for researchers, practitioners, and professional development leaders to explore modern and complementary computer-supported approaches for analysing and developing dialogue in classrooms. In this light, approaches that transcend human boundaries when studying classroom interactions in science will be explored. The specific focus is the automatic detection of teacher questions in the context of science. The focus on teacher questioning moves is rationalised by the essential role questions play in knowledge building and more engaging discursive activities developing reasoning and argumentation skills (O'Connor & Michaels, 2019; Sohmer et al., 2009).

Approaches to harness the power of automatic speech recognition (ASR) in analysis of content and semantic meaning (i.e., what is said) have been developed (Song et al., 2021); the next step is the automatic detection of pragmatic features (Nikula, 2005) of teacher talk (i.e., how it is said). In this study, the latter is addressed by targeting the focus on teacher questioning within the context of science classrooms. The theoretical background is foregrounded by sociocultural (Vygotsky, 1978) and dialogic (Bakhtin, 1986) perspectives highlighting the essentiality of social interactions. There is some evidence stressing that how scientific concepts are introduced in relation to one another makes a difference in terms of student learning (Viiri & Helaakoski, 2014). As shown later through more automated approaches, these results align with the results of automatic network analysis (Caballero et al., 2017). Similarly, an automatic keyword centrality approach—where three centrality measures were used to describe teachers' discourse—conformed to these findings (Schlotterbeck et al., 2020). These automated approaches made it possible to automatically identify differences between teachers by studying the way the physics concepts are presented during the lesson. By applying these measures, a new way of describing the content of teachers' talk was then introduced. In general, these findings underline the cruciality of the way the concepts are introduced and discussed in subject disciplines, such as science, as part of knowledge building (Barreto et al., 2021; Viiri & Helaakoski, 2014). In other words, it makes a difference how facts and concepts are interconnected and connected in relation to one another in condition and time (Badreddine & Buty,

2011; Barreto et al., 2021) and what kind of communication is used in knowledge building (Bereiter & Scardamalia, 2014).

The effectiveness of the orchestration of science content and instructional dialogue can be evaluated through the detection of elements characterised in high-quality instructional support and dialogue. In high-quality instructional dialogue, teachers' questioning strategies play a key role, especially when it comes to initiating student verbal participation and reasoning (Pianta et al., 2012). As evidenced, whilst one important aspect is the content of talk, high-quality instructional dialogue is facilitated through talk with different purposes, intentions, and communicative functions (Mortimer & Scott, 2003). In light of this research, it is argued that different types of teacher questioning serve different purposes. High-quality interaction and instructional dialogue can be considered to be facilitated through questioning that supports engagement, and sustainment of extended dialogues within students' thinking is stimulated and elaborated further (Chin, 2007; Salmon & Barrera, 2021). It is not only the questions but also the teacher's follow-up turns and questions (Cullen, 2002) and student responses that build towards more accountable talk (Sohmer et al., 2009) and dialogic interactions (Alexander, 2006). This is enforced especially through openness and extended dialogues; teacher initiations can still be counted as key mediators of such interactions. In other words, while acknowledging the limitation of the sole focus on teacher talk ASR outputs, the teacher's role as an orchestrator of the challenging knowledge-building process in science is emphasised in this study (Eshach, 2010; Mortimer & Machado, 2002).

The first aim of this study is to develop approaches for automatic detection of teacher questions. Then, the questions are automatically categorised into closed and open types. After introducing the results of the automatic analysis, a more thorough discussion on implications for education and multidisciplinary educational research is provided. By doing this, future possibilities for automatic analysis of classroom talk and the implications for teaching and learning are foregrounded. Whereas research on teaching and learning obviously deals with different fields and strands of research, the power of multidisciplinary approaches must still be harnessed to develop a merged approach bringing together the two sides—volume and accuracy. Before introducing the research questions and methods, an overview of the relevant theory on classroom interaction in science is presented, with the focus drawn to the role of teacher questioning.

Classroom Interaction in Science

The most fundamental form identified in the educational dialogue is the initiation-response-feedback (IRF) pattern of classroom talk (Sinclair & Coulthard, 1975). The triadic pattern is commonly orchestrated by a teacher and is related to a question-answer routine, and it has been found to dominate the interactions over the decades, especially in science (Lemke, 1990; Mercer et al., 2009). The nature of questions and teacher feedback could make a difference when heading towards more dialogic and extended exchange patterns (Lehesvuori et al., 2021; Ruiz-Primo, 2011; Scott et al., 2006). Within extended dialogues, questions are more open by nature to promote and stimulate students' thinking (Chin, 2007). Furthermore, teacher feedback/follow-up moves are not evaluative by nature (Cullen, 2002); rather, they include perpetuation (Bansal, 2018) or probing moves to engender ideation and elaborated reasoning. In this way, the pattern builds up to more of a chain kind of pattern (Scott et al., 2006). Extended dialogues have been linked to high-quality facilitation strategies of interaction and instructional dialogue (Pianta et al., 2012). Despite its effectiveness in designated situations, such as diagnostic questioning (Lehesvuori et al., 2022), breaking the dominant IRF pattern has the potential to increase dialogic interactions among students when delivered with new rules to engage in educational dialogue through openness and criticality (Gillies, 2016; Littleton & Mercer, 2013; Sohmer et al., 2009).

Communicative Approach Framework

The communicative approach framework (Mortimer & Scott, 2003) differentiates the typical transmission modes of teaching from classroom interaction, where students are given the freedom to, for example, describe, compare, classify, and argue when taking part in discussions. The framework consists of four categories generated from the combination of two dimensions: interactive/noninteractive and authoritative/dialogic. Interactive talk allows students to participate, whereas noninteractive talk is of a lecture type; additionally, whereas the dialogic approach considers diverging ideas, the authoritative approach focuses on a specific point of view, usually the scientific view, controlled by the teacher:

- In the question-answer routine of the interactive authoritative approach, the teacher often evaluates the students' responses and rejects divergent ideas. The authoritative approach focuses on the scientific point of view, and the questions are dominantly closed by nature.
- In marked contrast, within the interactive dialogic approach, the teacher elicits and explores students' ideas (e.g., everyday views) through more open questioning without the evaluative aspect. Thus, the dialogic approach is considered when the teacher is not trying to promote a specific point of view. Rather, the teacher tries to elicit the students' points of view and works with these different and even contrasting views.

- In the noninteractive authoritative approach, the teacher presents scientific content by presentation and takes no account of contrasting points of view.
- In the noninteractive dialogic approach, the teacher deals with contrasting points of view, such as students' everyday views, and shifts to present the scientific view. Thus, even though the teacher is lecturing, different ideas are discussed. Teacher presentations are therefore dialogic by nature.

In this study, when addressing the nature of teacher questions, indications of interactive communicative approaches will be perceived. Whereas closed questions indicate authoritative approaches, open questions might indicate a dialogic communicative approach (Lehesvuori et al., 2021). The triadic IRF pattern is often linked to the authoritative approach, and the dialogic approach is manifested in extended dialogues. The nature of the interaction and questioning could be evaluated through the level of interanimation. For example, despite being dialogic by nature, whereas simply collecting ideas aligns with a low level of interanimation, more conceptually loaded reasoning serves to increase interanimation (Mortimer & Scott, 2003).

The tension between authoritative and dialogic forms of interaction seems to be in-built in science classrooms (Author, 2019), which could be understood as the driving force of meaningful discussions (Scott & Ametller, 2007) within different forms of interaction seeding one another (Scott et al., 2006). Some preliminary evidence suggests that communication including both authoritative and dialogic approaches appear to be the most beneficial to student learning (Furtak & Shavelson, 2009). Aligning with Alexander's (2006) principles of cumulativeness and purposefulness, the importance of subject- and content-driven exchanges have been highlighted in classroom observation protocols for instructional dialogue (Hafen et al., 2015). However, a strict focus on content could hamper more authentic and dialogic forms of interaction as dialogic space is hindered through teachers directing the talk in their desired direction. The balance between the presentation of the content and dialogicity is challenging, yet there are some explicit solutions to open up dialogic spaces for student ideas and student-driven discussions. In light of this study, one approach would be to implement a more open questioning strategy.

Automatic Analysis of Classroom Interactions and Teacher Questioning

As some previous work on analysis of the content of teacher talk has already introduced, there are also other aspects that people have tried to capture through automatic approaches. For example, previous research has addressed detection of instructional activities (Donnelly et al., 2016; Lv et al., 2021; Wang et al., 2014), student roles in online group discussions (Dowell et al., 2019), and even emotional support and positive climate (Ramakrishnan et al., 2019). Specifically in the science context, automatization has been used to differentiate between phases of inquiry-based learning (Lämsä et al., 2021). When drawing on dialogicity and collectivity, technically, the diarisation (i.e., who speaks and when) could straightforwardly take place automatically from the audio (Kaseva et al., 2019). Human-composed transcriptions can also be used for this (cf. Lämsä et al., 2021). However, these approaches do not give information about teacher orchestration of classroom interaction in terms of finding the balance between content and forms of interaction. Some preliminary work has been done, for example, in relation to question-related keywords (e.g., "how" and "why"), which can be used in detection of questions, as well as their openness and authenticity (e.g., Blanchard et al., 2016; Huang et al., 2020; Kelly et al., 2018; Olney et al., 2017). In addition to the sentence, structural (Guyon & Elisseeff, 2003) and spectral features could lead to detection of questions and different question types.

Authentic questions can be assimilated with open questions to avoid answers being predefined by the teachers (Nystrand et al., 1997). The results on automatic detection of teacher questions are promising, yet one challenge comes along with the definition of open and authentic questions. Open questions are often related to higher-order cognitive activation, challenging, and furthering one's thinking (Bloom, 1956; Chin, 2007; Soter et al., 2008), but they can also serve the purpose of simply collecting student views without further interanimation of the ideas (Mortimer & Scott, 2003). Thus, consideration of the cognitive level of a question certainly brings in further challenges. Judgement of the function of a question often involves temporal consideration of the preceding and following interactions. One approach to investigations would be to seek indicators for questions and their types without further exploration of the surrounding dialogue. For example, whereas open questions could be indicators for dialogicity (Lehesvuori et al., 2021), some linguistic markers could be indicators for open questions. The developed classification models would be expected to try to process these indicators when computing selections for questions and question types.

Research Questions

In light of the presented literature and the targeted focus on teacher talk and questions, the following research questions will be addressed:

1. To what extent can potential questions be automatically detected from only ASR outputs of teacher talk?

2. To what extent can potential closed and open questions be differentiated from the detected questions?

The first question is about finding the possibility to detect questions simply from teacher talk and ASR outputs. The second question then investigates whether questions can be categorised into closed and open types. Based on the results, further suggestions on the interplay between manual and automatic analysis of classroom interactions will be discussed alongside educational implications.

METHODS

Data Collection and Analysis

The study data consists of 25 physics lessons from 25 different science teachers on the same topic of electricity and electrical energy, usually taught in the ninth grade in Finland. Each lesson was 90 minutes long. Lesson structuring was quite similar because teachers tend to lean on curriculum-compliant physics text- and workbooks when planning and teaching lessons. A typical lesson included activities such as checking homework, introducing the day's topic and key concepts, experimental work, and calculation tasks (Viiri & Helaakoski, 2014). Although the activities and executed tasks followed a typical pattern, differences have been found, especially when it comes to the orchestration of learning interactions (Lehesvuori et al., 2013). For example, teachers used an introductory textbook question ("When is someone being efficient?") very differently when building the link between the definitions of mechanical and electrical power. Whereas one teacher used this question to initiate a dialogic whole-class discussion, another teacher merely sought a brief and correct answer (Lehesvuori et al., 2021). All the lessons were video recorded with two video cameras, and separate microphones were placed each side of the classroom. In this study, the ASR output of the audio from a portable microphone carried by the teachers is used. The ASR output consists of the recognised speech divided into five-second intervals. In other words, all the words recognised within five seconds form a unit of words in a row (Table 1).

Manual annotations. When pursuing the automatic detection of teacher questions and question types, the first phase was to manually annotate all the explicit questions from the ASR outputs. In more detail, the following analytical human-led steps were implemented:

1. The researcher (Author 1) annotated all the indications of questions (i.e., potential questions) from the ASR outputs constructed from the teacher microphone audio. The annotation was based on detecting explicit questions.
2. The researcher decided whether the question was a closed or open question based on the theory-driven (Chin, 2007) categorisation (see Tables 1 and 2).
3. Based on the annotation process, hand-crafted baseline rules were developed based on steps 1 and 2 (see Table 2). These rules were used as a baseline for the automatic detection.
4. The researcher checked the reliability of the annotations against 20 per cent of the audio recorded by the teacher microphone. Similarly, the inter-rater agreement on the questions and question types was found to be sufficient, which is discussed in another study focusing solely on the manual aspects of the analysis (in preparation).

Table 1. Example of ASR text output annotations (code 1 = true)

Question	Closed question	Open question	Time	ASR text output (English translation added)
1		1	485-490	miksi kierroslukumäärä piti vaihtaa tai muuttaa (Why do the rounds need to be changed?)
			490-495	iha yksinkertanen selitys jotta (Kind of simple explanation, though)
1		1	495-500	apua <NAME REMOVED> auta minkä takia kierroslukumäärä mutta (Some help <NAME REMOVED>; why do the rounds need to be changed?)

Annotation Procedure and Cases of Exceptions

The accuracy of the developed ASR system was found to be sufficient for analysing the content (Sclotterbeck et al., 2021a) and providing transcription assistance (Caballero et al., 2017). One should note that Finnish is a small language in terms of people speaking or writing it; thus, a context-language-specific system was developed in collaboration with leading academic experts in the field (Kronholm et al., 2017). Although some words were missing or mistyped, the human coder was generally able to annotate the questions. However, if a question word was completely missing, it was difficult to detect a question. These missed cases were confirmed to be very few after validating the human ASR output annotations against the original audio file.

Annotation was placed in the five-second row including the actual question word or sentence (e.g., "how" or "do you remember?"). In other words, if a questioning sentence was located in several rows (usually a maximum of two), the annotation was placed on the row with the explicit questioning word. A new question in close temporal surroundings was annotated when a teacher was asking for follow-up from another student (e.g., "Laura?") or seeking elaboration (e.g., "Can you open up this a bit more?"). That is, prompting subsequent questions were all annotated; thus, there could be several codes in neighbouring rows. Although not explored here, these chained questions are also interesting and could indicate more open and extended dialogues. A question was not coded when the question was simply and clearly a part of teacher lecturing ("We cannot say what this is"). If a question was immediately reformulated without changing the nature of the question, then it was not coded twice (e.g., "What is this? I mean, what are these things here?"). A question was also not coded when a teacher responded to the question him or herself as a part of the teacher presentation ("So, what we have here is a situation with two lamps in parallel…"). Additionally, sarcastic and/or rhetorical questions controlling student behaviour were not coded (e.g., "Are you having fun?"). Finally, question-like suggestions were not coded (e.g., "Could you turn off the lights, please?"). Briefly, a question was considered when it was related to the content and execution of the experiments and inquiries. As expected, although the human annotator had a run-through of the automatic detection results, some of these instances were coded as questions by the developed system.

Table 2. Descriptions for annotating potential questions and question types

Question	Closed question	Open question
A question was annotated to ASR output when it was explicitly indicated. Explicit questions were often indicated through a questioning word or phrase (e.g., "why", "how", "where", "how many", "what is").	A closed question seeks correct answers or there is little flexibility to formulate a response. A closed question is indicated, for example, by questions seeking facts, remembering, or certain numeric values for calculations (e.g., "What is the unit of power?"; "How much is this?").	An open question seeks elaboration of ideas or different opinions and voices. Open questions may stimulate reasoning (Chin, 2007), or the teacher might simply be collecting ideas and/ or guesses (e.g., "What kind of electrical devices do you have back home?"). Open questions were indicated often by certain keywords and sentences (e.g., "why", "how", "What do you think", "Go ahead, what is your guess").
Hand-crafted rule example: If at least one word from set A {how, what, when…} appears, then the segment is a question; otherwise, it is not a question.	*Hand-crafted rule example:* If at least one word from set A or B {how much, how many, remember…} appears and one word from set C {get, comes…} or D {unit, precision…} appears, then the segment indicates a closed question.	*Hand-crafted rule example:* If a word from set A and one word from set E {home, at home…} or F {think about, guess…} appears, then the segment indicates an open question.

As has been found, dialogic discussions are very seldom in these classes (Lehesvuori et al., 2013); it is also expected that closed questions are the dominant type. Thus, it would not be worthwhile to categorise the questions here into further subcategories, such as based on Bloom's (1956) taxonomy. That is, to ensure a sufficient balance between the occurrences of different types of questions, the research retains a simple division between closed and open ones. It should be noted that final judgements cannot be made on whether a question is closed or open without considering student turns and teacher follow-ups (Graesser et al., 1992). However, whether a question is indicatively closed or open by nature can be determined.

Automated Features

Hand-Crafted Rules

First, a set of hand-crafted rules of the form "if n or more words from a predefined set A appear in a sentence, then that sentence is (not) a (open/closed) question" was developed to be used as a baseline. In total, eight rules were implemented (see examples in Table 2), the first three indicating whether the sentence is a question, and the other five indicating if it is an open or closed question. These rules were then applied to each lesson, and the one that yielded the best average balanced accuracy across all lessons was finally selected as the baseline for each respective task.

Aside from the hand-crafted rules, two machine learning models were used to extract linguistic features from the transcripts: a Bag of Words (BoW) and a BERT-based approach. These models assign to each sentence a high-dimensional vector representation, which can be forwarded as the input for the two classification models. Besides running these models alone, they were also combined with the hand-crafted rules to potentially enhance the performance. Brief descriptions of these models are presented below.

Bag of Words Model

The first approach considered for the feature extraction was a BoW model. This model represents a text sequence, such as a sentence or a document, by counting the number of times a particular word or tuple of words (known as an N-gram) appears in the sequence. Thus, given that each N-gram has its own dimension on the vector, an advantage of this approach lies in its interpretability and sparsity. In this way, it is possible to identify pivotal words based on the coefficients associated in the classifier, given that the coefficients in the classifier are interpretable, as is the case in our experiments. Moreover, the list of words found during experimentation was evaluated by the researcher who conducted the manual annotation and was found to consist of words that are relevant to the tasks set for the classification models.

BERT model

The second approach used to extract linguistic features was the BERT-based one. BERT is a natural language processing (NLP) model developed in 2018 by Google and has been the state of the art for various NLP tasks since then. This is due to the incorporation of the transfer learning approach in which the model learns high-dimensional, contextualised representations of words using fill-in-the-blank questions. Later, these representations can be adjusted and used for more downstream tasks such as text classification or next-sentence prediction. This tandem of procedures is usually known as pretraining and fine-tuning the model, respectively. During experimentation, the full high-dimensional representations were extracted for every word while assigning to each sentence the sum of its respective words. The experimentation was carried out using the large version of ProsusAI's Finnish BERT. The language-specific model can be applied to achieve state-of-the-art results for different Finnish NLP tasks (Virtanen et al., 2019).

Classification Models to Detect Questions

Once the feature extraction process was finished, the obtained representations were forwarded to a classification model to solve two classification tasks. The first task was to determine whether the unit (5-second row) was a question. The second task was to determine whether a question was open or closed. This means that a dichotomous and binary classification was applied to each task (a value of 0 equals false; a value of 1 equals true).

Two options were considered for the classification model: logistic regression and gradient-boosted trees. The first one classifies the inputs by applying a one-dimensional affine transformation and then a sigmoid function, returning a number between 0 and 1, which can be interpreted as the probability that the given input is from the positive class. Next, the regression coefficients are optimised by minimising the logistic loss between the predicted probabilities and the ground truth classes. On the other hand, the latter uses an iteratively generated ensemble of decision trees such that in each iteration, the tree is fit over the remaining errors from the past iteration. Therefore, the predictive capability of the model is enhanced with each classifier added, achieving better results than the random forest approach.

The classifiers were first trained using the annotations of the manual analysis as targets and only the extracted features as input. To avoid overfitting, regularisation terms were added to the loss function of each classifier, using l2 and l1 norms for the BERT- and BoW-based approaches, respectively, considering the sparsity of the BoW representation. Additionally, to estimate the generalisation capability of the classifiers, the models were trained and assessed using a leave-one-out cross-validation approach across the lessons. On each fold, the predictive quality of the models was assessed using four common evaluation metrics for classification tasks, which is briefly described next:

- *Precision* is the ratio between the number of correctly classified positives (question/open question) and the total number of utterances that our model classified as positives (i.e., out of the ones that the model deemed questions, what proportion corresponded to real questions?).
- *Recall* is the ratio between the number of correctly classified positives and the total number of real positives (i.e., out of the ones that were real positives, how many did we identify?).
- *Balanced accuracy* is the average of the recalls obtained for the positive and negative classes (i.e., 0.5*(recall(isQuestion) + recall(isNonQuestion))).
- Finally, *F1 score* is the harmonic mean between precision and recall.

Lastly, as a result of the iteration process and test runs, a combined-feature approach was developed in which the classifiers were input with the concatenation of the results from the hand-crafted rules and the automated features. Figure 1 presents a summary of the development procedure.

Figure 1. The development of the automatization and classification models

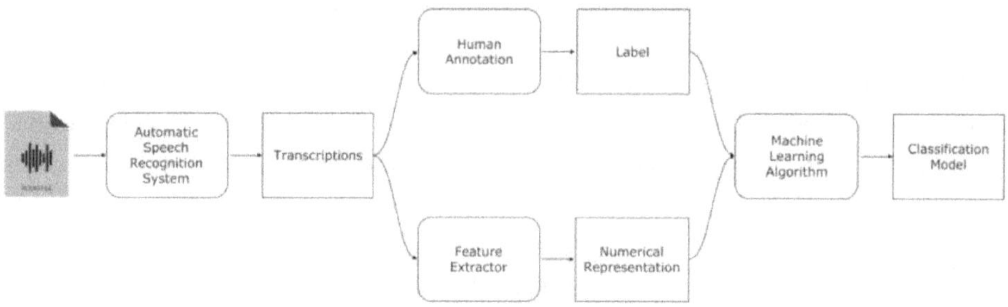

RESULTS

In the test setting, the hand-crafted rules were compared with four introduced machine learning methods for two tasks. Table 3 summarises the results obtained with automated features and related classification models (see left column). In general, the greater the numeric value was, the greater the performance was for the tasks. The evaluation metrics were described earlier and are reviewed briefly in Table 3.

Table 3. The precision, recall, balanced accuracy and F1 score for detecting questions and question types: Comparison of hand-crafted and automated approaches

Task		Question vs Non-question				Open vs Closed			
Features	**Classif.**	**P***	**R****	**BA*****	**F1******	**P**	**R**	**BA**	**F1**
HCR	-	0.351	0.609	0.77	0.436	0.269	0.994	0.514	0.411
BoW	**LR**	0.185	0.529	0.695	0.27	0.511	0.553	0.685	0.51
	GBT	0.172	0.69	0.744	0.272	0.522	0.532	0.679	0.512
BERT	**LR**	0.131	0.567	0.669	0.211	0.392	0.527	0.622	0.438
	GBT	0.13	0.517	0.652	0.205	0.368	0.355	0.57	0.348
BoW + HCR	**LR**	0.298	0.7	0.799	0.411	0.496	0.542	0.677	0.498
	GBT	0.311	0.742	0.821	0.431	0.52	0.521	0.679	0.503
BERT + HCR	**LR**	0.149	0.615	0.7	0.237	0.385	0.523	0.626	0.423
	GBT	0.164	0.71	0.745	0.263	0.368	0.355	0.57	0.348

BERT + HCR = combination of the BERT model and hand-crafted rule; BoW = Bag of Words; BoW + HCR = combination of the Bag of Words model and hand-crafted rule; GBT = gradient-boosted trees; HCR = hand-crafted rule; LR = logistic regression.

*Ratio between the number of correctly classified positives (question/open question) and the total number of utterances that our model classified as positives.

**Ratio between the number of correctly classified positives and the total number of real positives.

***Balanced accuracy is the average of the recalls obtained for the positive and negative classes.

****The harmonic mean between precision and recall.

In the task to determine if the segment was a question, the best performance was obtained using the hand-crafted rules, in terms of F1 score. If the focus of a system is based on recall, the best model is the one that combines BoW features with the hand-crafted rules. All this indicates that the presence of particular terms is critical to identify questions. On the other hand, in the task of distinguishing between open and closed questions, the best F1 score was obtained using a BoW representation, showing that there is enough data to identify relevant terms in the sentences that help to distinguish between open and closed questions. However, it can also be noted that the hand-crafted rules have a big recall compared with the explored machine learning approaches, suggesting that the hand-crafted rules provide a necessary condition to differentiate between open and closed questions. It is also important to note that the hand-crafted rules have the lowest precision value, indicating that this approach is far from providing sufficient conditions to complete this task.

Overall, the results show the feasibility of developing systems that can detect questions and types of questions from ASR output with good performance, understanding that there is room to continue improving these metrics.

DISCUSSION

It is generally agreed that researchers should take advantage of emerging techniques to reduce the time required for coding classroom activities (see, e.g., Chan et al., 2020). The aim of this study was to find out whether and to what extent teacher questions could be traced automatically merely from ASR outputs. A follow-up challenge was then to categorise the questions as potentially closed and open by nature. The results of this study suggest that questions can be detected to a moderate extent using the basic models, whereas with the combination models, the results clearly improved. The results in relation to the follow-up task showed that both automated approaches clearly outperformed solely hand-crafted baseline rules when categorising questions as closed and open types. This is an interesting finding and requires further investigations to delve into the reasons behind it. One key to success might be the context-specific ASR outputs consisting of the same lessons delivered by different teachers. This could mean that in addition to common question-related keywords, frequent lesson-specific keywords might be associated, for example, with open questioning. An example of such a non-question- and non-content-specific keyword is the word "home", which in many cases was linked to dialogue by which a teacher was seeking student ideas about domestic electrical devices during an idea-collecting activity. Connecting this with the theoretical background, this kind of interaction resembles a dialogic approach with low interanimation of ideas (Mortimer & Scott, 2003), which has been discussed as the easiest approach for science teachers to engage in dialogic discussions (Lehesvuori et al., 2011).

Automated approaches could provide easy access to study how teachers orchestrate classroom interactions and potential dialogues in science classrooms. The existence of open questions could indicate more dialogic communicative approaches and dialogic teaching strategies being implemented in the classroom. Moreover, open questions with a higher level of interanimation could initiate higher-level cognitive engagement, which is expected to affect student learning (Ergönenç et al., 2014). Although the cognitive level of the questions (e.g., Bloom, 1956) was not measured in this study, the open and closed categorisation can provide essential information about how communication is orchestrated in terms of meaningful learning. In other words, although dialogicity and open questions are stressed due to their infrequency, meaningful science learning also involves authoritative approaches, often indicated through more closed, targeted, and diagnostic questions. This is a fundamental part of meaningful learning within one communicative approach that gives rise to another (Scott et al., 2006). The changes in communicative approaches could be traced in changes of patterns of interactions or linguistic markers (cf. Lehesvuori et al., 2013), which in this study's case could be based on the existence and changes in the types of questions. Specific to science instruction, the use of different communicative approaches plays a role in learning when dealing with and moving between different representations of scientific information, as well as student ideas and everyday views (Mortimer & Scott, 2003).

Keeping in mind that these results are mainly based on the use of ASR outputs, the automatization procedure could be strengthened by bringing in the audio data (cf. Kelly et al., 2018). It is also obvious that other features, such as visual cues and proxemics, would influence real observers' decisions about classifications. It is clear that automated methods cannot achieve the accuracy of human coding based on highly developed and validated observational protocols (Pianta et al., 2012), but they can provide the initial filtering tool for detecting classrooms with potentially richer dialogue. Indeed, it can be argued that some indicator-level features of a higher level of instructional dialogue, including open questions, active listening, and supportive feedback, could be detected in multimodal data (Lehesvuori et al., 2021). Additionally, when enough of this kind of information is combined, it would be possible to build models to help go from the utterance level back to the observation and episode level of analysis, which is something that an educational researcher often implicitly does during observations.

Despite manual coding features such as classroom activities, forms of instruction, and question types, many questions remain unanswered when it comes to the use of language (Fischer et al., 2014). Further elaborations on multidisciplinary work will surely bring in some answers in relation to this matter via more effective use of the data. In light of this study, detecting indications of more open-ended questions is essential when heading towards more student-centred and dialogic approaches. Some indications of the use of more dialogic strategies and their association with student learning have been discussed (Muhonen et al., 2018), yet further research is needed to investigate this connection.

IMPLICATIONS AND LIMITATIONS

ASR outputs of teacher talk were found to be sufficient to detect types of questions at the indicative level. However, judging whether students see these indicatives as open or closed calls for consideration of multimodal data, such as audio (e.g., tone of speech and intonation) and video data (e.g., gestures and proxemics). Most importantly, to evaluate the effect of the indicatives on the joint construction of educational dialogue, students' voices need to be considered. One straightforward possibility would be automatic detection of student wonderment questions, which is essential when studying more dialogic communicative approaches (Aguiar et al., 2009). Despite the acknowledged limitations of missing multimodal information (Kääntä, 2015), dealing with ASR outputs eliminates many ethical risks of handling the data, for example, in an international study such as that reported within this chapter. When bringing in this multimodal information, future researchers could delve into exploring teacher communicative approaches based on the questioning strategies. In addition, a class-level consideration of learning gains and their association with questioning could be investigated.

Even though the results show that using the introduced models could be a promising starting point, there are some further limitations to be considered. For example, the ASR outputs are some years old, and it can be expected that newer versions of the ASR system would yield more precise data to analyse. Despite the limited data, recent research has shown that it is possible to train an ASR system adapted to classroom environments to achieve good performance (Schlotterbeck et al., 2022). Furthermore, the proportion of questions in the whole corpus is somewhat small. Thus, future work in multimodal analysis, which involves the use of different sources of data, could help to improve the metrics of the proposed models with richer classroom data. Although the data seems sufficient for educational researchers, for computational and machine learning purposes, it is very limited. However, this was also the driving force for the intensive interdisciplinary collaboration within the development of the classification models, which was conducted in weekly interplay with researchers from different strands. Elaborating this further, whereas educational research deals with the laboriousness of analysis, computing experts would be interested in authentic and real data. Despite the limitations pointed out, the process surely was a step towards bridging the gap between accuracy and effectiveness. Continuation of this work could lead to enhanced ability to harness computational approaches for the support of classroom interaction research purposes, which would magnify the usability of the data collected in related studies in the future.

Implications for education and teacher education can address how teachers can enrich their questioning strategies via feedback enabled through automatic detection of desired features. Open questions have been shown to be infrequent, especially in science classrooms (Lehesvuori et al., 2013), but their use can be augmented through teacher education and increased awareness (Oliveira, 2010). Thus, it would be worthwhile to point this out to the teachers. Compared with manual approaches, a low-cost and less time-consuming approach was presented in this study. The developed models can provide valuable feedback to teachers so they can reflect on their own practice without having to go through the whole lesson to extract the times when they posed questions to students. For example, whereas videos have been shown to be powerful tools to reflect one's educational practice (Borko et al., 2014), it is very time consuming to watch them, and thus, a common teacher might not engage in such a practice. Therefore, even indications of detected and reflected practice could save time and resources. In this way, classifications of teacher discourse, alongside the automatic recognition of questions and question types, would provide immediate support to researchers and teachers (cf. Jacobs et al., 2022). Finally, in tune with the fourth industrial revolution, a step forward was taken by harnessing ASR and machine learning to analyse classroom interactions in science.

ACKNOWLEDGMENT

Support from ANID/PIA/Basal Funds for Centers of Excellence FB0003 is gratefully acknowledged.

REFERENCES

Aguiar, O. G., Mortimer, E. F., & Scott, P. (2009). Learning from and responding to students' questions: The authoritative and dialogic tension. *Journal of Research in Science Teaching*, *47*(2), 174–193. doi:10.1002/tea.20315

Alexander, R. (2006). *Towards dialogic teaching* (3rd ed.). Dialogos.

Badreddine, Z., & Buty, C. (2011). Discursive reconstruction of the scientific story in a teaching sequence. *International Journal of Science Education*, *33*(6), 773–795. doi:10.1080/09500693.2010.496475

Bakhtin, M. M. (1986). *The dialogic imagination: Four essays (M. Holquist* (C. Emerson & M. Holquist Eds. & Trans.). University of Texas Press.

Bansal, G. (2018). Teacher discursive moves: Conceptualising a schema of dialogic discourse in science classrooms. *International Journal of Science Education*, *40*(15), 1891–1912. doi:10.1080/09500693.2018.1514543

Barreto, L. P., Rodrigues, A. A. D., Oliveira, G. C. B., Almeida, L. T. G., Felix, M. A. C., Silva, P. S., Quadros, A. L., Macedo, A. M., & Mortimer, E. F. (2021). The use of different translation devices to analyze knowledge-building in a university chemistry classroom. *Research in Science Education*, *51*(1), 135–152. doi:10.100711165-020-09969-z

Bereiter, C., & Scardamalia, M. (2014). Knowledge building and knowledge creation: One concept, two hills to climb. In S. C. Tan, H. J. So, & J. Yeo (Eds.), *Knowledge creation in education* (pp. 35–52). Springer. doi:10.1007/978-981-287-047-6_3

Blanchard, N., Donnelly, P. J., Olney, A. M., Samei, B., Ward, B., Sun, X., Kelly, S., Nystrand, M., & D'Mello, S. K. (2016). Semi-automatic detection of teacher questions from human-transcripts of audio in live classrooms. *Proceedings of the 9th International Conference on Educational Data Mining (EDM 2016)* (pp. 288-291). Springer.

Bloom, B. S. (1956). *Taxonomy of educational objectives, handbook: The cognitive domain*. David McKay.

Borko, H., Jacobs, J., Seago, N., & Mangram, C. (2014). Facilitating videobased professional development: Planning and orchestrating productive discussions. In Y. Li, E. A. Silver, & S. Li (Eds.), *Transforming mathematics instruction: Multiple approaches and practices* (pp. 259–281). Springer. doi:10.1007/978-3-319-04993-9_16

Caballero, D., Araya, R., Kronholm, H., Viiri, J., Mansikkaniemi, A., Lehesvuori, S., Virtanen, T., & Kurimo, M. (2017). ASR in Classroom Today: Automatic Visualization of Conceptual Network in Science Classrooms. In É. Lavoué, H. Drachsler, K. Verbert, J. Broisin, & M. Pérez-Sanagustín (Eds.), *Data Driven Approaches in Digital Education: 12th European Conference on Technology Enhanced Learning, EC-TEL 2017,* (pp. 541-544). Springer International Publishing. 10.1007/978-3-319-66610-5_58

Chan, M. C. E., Ochoa, X., & Clarke, D. (2020). Multimodal Learning Analytics in a Laboratory Classroom. In M. Virvou, E. Alepis, G. Tsihrintzis, & L. Jain (Eds.), *Machine Learning Paradigms. Intelligent Systems Reference Library* (Vol. 158). Springer. doi:10.1007/978-3-030-13743-4_8

Chin, C. (2007). Teacher questioning in science classrooms: Approaches that stimulate productive thinking. *Journal of Research in Science Teaching*, *44*(6), 815–843. doi:10.1002/tea.20171

Cullen, R. (2002). Supportive teacher talk: The importance of the F-move. *ELT Journal*, *56*(2), 117–126. doi:10.1093/elt/56.2.117

Donnelly, P. J., Blanchard, N., Samei, B., Olney, A., Sun, X., Ward, B., Kelly, S., Nystrand, M., & D'Mello, S. K. (2016). Automatic Teacher Modeling from Live Classroom Audio. *UMAP '16: Proceedings of the 2016 Conference on User Modeling Adaptation and Personalization*. ACM. 10.1145/2930238.2930250

Dowell, N. M. M., Nixon, T. M., & Graesser, A. C. (2019). Group communication analysis: A computational linguistics approach for detecting sociocognitive roles in multiparty interactions. *Behavior Research Methods*, *51*(3), 1007–1041. doi:10.375813428-018-1102-z PMID:30120683

Ergönenç, J., Neumann, K., & Fischer, H. E. (2014). The impact of pedagogical content knowledge on cognitive activation and student learning. In H. E. Fischer, P. Labudde, K. Neumann, & J. Viiri (Eds.), *Quality of instruction in physics: Comparing Finland*. Waxmann Verlag.

Eshach, H. (2010). An analysis of conceptual flow patterns and structures in the physics classroom. *International Journal of Science Education*, *32*(4), 451–477. doi:10.1080/09500690802635247

Fischer, H. E., Neumann, K., Labudde, P., & Viiri, J. (Eds.). (2014). *Quality of instruction in physics: Comparing Finland*. Waxmann Verlag.

Furtak, E. M., & Shavelson, R. J. (2009). Guidance, Conceptual Understanding, and Student Learning: An Investigation of Inquiry-Based Teaching in the US. In T. Janik & T. Seidel (Eds.), *The Power of Video Studies in Investigating Teaching and Learning in the Classroom* (pp. 181–206). Waxmann.

Geller, C., Neumann, K., Boone, W. J., & Fischer, H. E. (2014). What makes the Finnish different in science? Assessing and comparing students' science learning in three countries. *International Journal of Science Education*, *36*(18), 3042–3066. doi:10.1080/09500693.2014.950185

Gillies, R. (2016). Dialogic interactions in the cooperative classrooms. *International Journal of Educational Research*, *76*, 178–189. doi:10.1016/j.ijer.2015.02.009

Graesser, A. C., Person, N. K., & Huber, J. D. (1992). Mechanisms that generate questions. In T. E. Lauer, E. Peacock, & A. C. Graesser (Eds.), *Questions and information systems* (pp. 67–187). Erlbaum.

Guyon, I., & Elisseeff, A. (2003). An introduction to variable and feature selection. *Journal of Machine Learning Research*, *3*, 1157–1182.

Hafen, C. A., Hamre, B. K., Allen, J. P., Bell, C. A., Gitomer, D. H., & Pianta, R. C. (2015). Teaching through interactions in secondary school classrooms: Revisiting the factor structure and practical application of the classroom assessment scoring system–Secondary. *The Journal of Early Adolescence*, *35*(5–6), 651–680. doi:10.1177/0272431614537117 PMID:28232770

Hennessy, S., Howe, C., Mercer, N., & Vrikki, M. (2020). Coding classroom dialogue: Methodological considerations for researchers. *Learning, Culture and Social Interaction*, *25*, 100404. doi:10.1016/j.lcsi.2020.100404

Howe, C., Hennessy, S., Mercer, N., Vrikki, M., & Wheatley, L. (2019). Teacher–student dialogue during classroom teaching: Does it really impact on student outcomes? *Journal of the Learning Sciences*, *28*(4-5), 462–512. doi:10.1080/10508406.2019.1573730

Huang, G. Y., Chen, J., Liu, H., Fu, W., Ding, W., Tang, J., Yang, S., Li, G., & Liu, Z. (2020). Neural multi-task learning for teacher question detection in online classrooms. https://arxiv.org/abs/2005.07845v1 doi:10.1007/978-3-030-52237-7_22

Jacobs, J., Scornavacco, K., Harty, C., Suresh, A., Lai, V., & Sumner, T. (2022). Promoting rich discussions in mathematics classrooms: Using personalized, automated feedback to support reflection and instructional change. *Teaching and Teacher Education*, *112*, 103631. doi:10.1016/j.tate.2022.103631

Kääntä, L. (2015). The multimodal organization of teacher-led classroom interaction. In C. Jenks & P. Seedhouse (Eds.), *International perspectives on ELT classroom interaction* (pp. 64–83). Palgrave Macmillan. doi:10.1057/9781137340733_5

Kaseva, T., Rouhe, A., & Kurimo, M. (2019). SphereDiar: An effective speaker diarization system for meeting data. Annual Conference of the International Speech Communication Association. ISCA.

Kelly, S., Olney, A. M., Donnelly, P., Nystrand, M., & D'Mello, S. K. (2018). Automatically measuring question authenticity in real-world classrooms. *Educational Researcher*, *47*(7), 451–464. doi:10.3102/0013189X18785613

Lämsä, J., Uribe, P., Jiménez, A., Caballero, D., Hämäläinen, R., & Araya, R. (2021). Deep Networks for Collaboration Analytics: Promoting Automatic Analysis of Face-to-Face Interaction in the Context of Inquiry-Based Learning. *Journal of Learning Analytics*, *8*(1), 113–125. doi:10.18608/jla.2021.7118

Lehesvuori, S., & Ametller, J. (2021). Exploring coherence and authorship in pedagogical link-making in science. *International Journal of Science Education*, *43*(17), 2791–2813. doi:10.1080/09500693.2021.1991599

Lehesvuori, S., Ketonen, L., & Hähkiöniemi, M. (2022). Utilizing Informal Formative Assessment and Dialogicity During Reflections on Educational Dialogue in Mathematics. *Studia Paedagogica*, *27*(2), 55–75. doi:10.5817/SP2022-2-3

Lehesvuori, S., Viiri, J., & Rasku-Puttonen, H. (2011). Introducing Dialogic Teaching to Science Student Teachers. *Journal of Science Teacher Education*, *22*(8), 705–727. doi:10.100710972-011-9253-0

Lehesvuori, S., Viiri, J., Rasku-Puttonen, H., Moate, J., & Helaakoski, J. (2013). Visualizing communication structures in science classrooms: Tracing cumulativity in teacher-led whole class discussions. *Journal of Research in Science Teaching*, *50*(8), 912–939. doi:10.1002/tea.21100

Lemke, J. L. (1990). *Talking science: Language, learning and values*. Ablex Publishing Company.

Littleton, K., & Mercer, N. (2013). *Interthinking: Putting talk to work*. Routledge. doi:10.4324/9780203809433

Lv, K., Sun, Z., & Xu, M. (2021). Artificial Intelligent Based Video Analysis on the Teaching Interaction Patterns in Classroom Environment. *International Journal of Information and Education Technology (IJIET)*, *11*(3), 126–130. doi:10.18178/ijiet.2021.11.3.1500

Mercer, N., Dawes, L., & Staarman, J. K. (2009). Dialogic teaching in the primary science classroom. *Language and Education*, *23*(4), 353–369. doi:10.1080/09500780902954273

Mortimer, E. F., & Machado, A. H. (2000). Anomalies and conflicts in classroom discourse. *Science Education*, *84*(4), 429–444. doi:10.1002/1098-237X(200007)84:4<429::AID-SCE1>3.0.CO;2-#

Mortimer, E. F., & Scott, P. (2003). *Meaning making in science classrooms*. Open University Press.

Muhonen, H., Pakarinen, E., Poikkeus, A.-M., Lerkkanen, M.-K., & Rasku-Puttonen, H. (2018). Quality of educational dialogue and association with students' academic performance. *Learning and Instruction*, *55*, 67–79. doi:10.1016/j.learninstruc.2017.09.007

Müller, K., Prenzel, M., Seidel, T., Schiepe-Tiska, A., & Kjærnsli, M. (2016). Science teaching and learning in schools: Theoretical and empirical foundations for investigating classroom-level processes. In S. Kuger, E. Klieme, N. Jude, & D. Kaplan (Eds.), *Assessing contexts of learning, methodology of educational measurement and assessment* (pp. 423–446). Springer. doi:10.1007/978-3-319-45357-6_17

Nikula, T. (2005). English as an object and tool of study in classrooms: Interactional effects and pragmatic implications. *Linguistics and Education*, *16*(1), 27–58. doi:10.1016/j.linged.2005.10.001

Nystrand, M., Gamoran, A., Kachur, R., & Prendergast, C. (1997). *Opening dialogue: Understanding the dynamics of language and learning in the English classroom.* Teachers College Press.

O'Connor, C., & Michaels, S. (2019). Supporting teachers in taking up productive talk moves: The long road to professional learning at scale. *International Journal of Educational Research, 97*, 166–175. doi:10.1016/j.ijer.2017.11.003

Oliveira, A. W. (2010). Improving teacher questioning in science inquiry discussions through professional development. *Journal of Research in Science Teaching, 447*(4), 422–453. doi:10.1002/tea.20345

Olney, A. M., Kelly, S., Samei, B., Donnelly, P., & D'Mello, S. K. (2017). Assessing teacher questions in classrooms. In R. Sottilaire, A. Graesser, X. Hu, & G. Goodwin (Eds.), *Design recommendations for intelligent tutoring systems: Assessment* (Vol. 5, pp. 261–274). U.S. Army Research Laboratory.

Pianta, R. C., Hamre, B. K., & Mintz, S. (2012). *Classroom assessment scoring system – Secondary manual.* Teachstone.

Qader, W., Ameen, M., & Ahmed, B. (2019). *An overview of Bag of Words; Importance, implementation, applications, and challenges. International Engineering Conference.* 10.1109/IEC47844.2019.8950616

Ramakrishnan, A., Ottmar, E., LoCasale-Crouch, J., & Whitehill, J. (2019). Toward automated classroom observation: Predicting positive and negative climate. *14th IEEE Int. Conf. on Automatic Face and Gesture Recognition (FG 2019).* IEEE. 10.1109/FG.2019.8756529

Ruiz-Primo, M. A. (2011). Informal formative assessment: The role of instructional dialogues in assessing students' learning. *Studies in Educational Evaluation, 37*(1), 15–24. doi:10.1016/j.stueduc.2011.04.003

Salmon, A. K., & Barrera, M. X. (2021). Intentional questioning to promote thinking and learning. *Thinking Skills and Creativity, 40*, 100822. doi:10.1016/j.tsc.2021.100822

Schlotterbeck, D., Araya, R., Caballero, D., Jiménez, A., Lehesvuori, S., & Viiri, J. (2020). Assessing Teacher's Discourse Effect on Students' Learning: A Keyword Centrality Approach. In C. Alario-Hoyos, M. J. Rodríguez-Triana, M. Scheffel, I. Arnedillo-Sánchez, & S. M. Dennerlein (Eds.), EC-TEL 2020: Addressing Global Challenges and Quality Education (pp. 102-116).

Schlotterbeck, D., Jiménez, A., Araya, R., Caballero, D., Uribe, P., & Van der Molen Moris, J. (2022). "Teacher, Can You Say It Again?" Improving Automatic Speech Recognition Performance over Classroom Environments with Limited Data. In M. M. Rodrigo, N. Matsuda, A. I. Cristea, & V. Dimitrova (Eds.), Lecture Notes in Computer Science: *Artificial Intelligence in Education. AIED 2022*. Springer. doi:10.1007/978-3-031-11644-5_22

Scott, P. H., Mortimer, E. F., & Aguiar, D. G. (2006). The tension between authoritative and dialogic discourse: A fundamental characteristic of meaning making interactions in high school science lessons. *Science Education*, *90*(3), 605–631. doi:10.1002ce.20131

Sedlacek, M., & Sedova, K. (2017). How many are talking? The role of collectivity in dialogic teaching. *International Journal of Educational Research*, *85*, 99–108. doi:10.1016/j.ijer.2017.07.001

Sedova, K. (2017). A case study of a transition to dialogic teaching as a process of gradual change. *Teaching and Teacher Education*, *67*, 278–290. doi:10.1016/j.tate.2017.06.018

Sinclair, J., & Coulthard, R. M. (1975). *Towards an analysis of discourse*. Oxford University Press.

Sohmer, R., Michaels, S., O'Connor, M. C., & Resnick, L. (2009). Guided construction of knowledge in the classroom: The troika of talk, tasks and tools. In B. Schwarz, T. Dreyfus, & R. Hershkowitz (Eds.), *Transformation of knowledge through classroom interaction* (pp. 105–129). Routledge.

Song, Y., Lei, S., Hao, T., Lan, Z., & Ding, Y. (2021). Automatic classification of semantic content of classroom dialogue. *Journal of Educational Computing Research*, *59*(3), 496–521. doi:10.1177/0735633120968554

Soter, A. O., Wilkinson, I. A. G., Murphy, P. K., Rudge, L., Reninger, K., & Edwards, M. (2008). What the discourse tells us: Talk and indicators of high-level comprehension. *International Journal of Educational Research*, *47*(6), 372–391. doi:10.1016/j.ijer.2009.01.001

Viiri, J., & Helaakoski, J. (2014). Content and content structure of physics lessons and students' learning gains: Comparing Finland, Germany and Switzerland. In H. E. Fischer, P. Labudde, K. Neumann, & J. Viiri (Eds.), Quality of instruction in physics: Comparing Finland, Germany and Switzerland (pp. 93–110). Waxmann Verlag.

Virtanen, A., Kanerva, J., Ilo, R., Luoma, J., Luotolahti, J., Salakoski, T., Ginter, F., & Pyysalo, S. (2019). Multilingual is not enough: BERT for Finnish. arXiv preprint arXiv:1912.07076

Vygotsky, L. S. (1978). *Mind in society: The development of higher psychological processes*. Harvard University Press.

KEY TERMS AND DEFINITIONS

Automated Features: Machine learning models that can be used to extract linguistic features from the transcripts.

Automatic Speech Recognition System: In this study: detection of speech formulated as a text file.

Classification Models: Can be used to solve two classification tasks: such as classifying a question as open or not by nature.

Closed Questions: Questions that seek predefined answers or remembering, or have little flexibility for student-formulated responses.

Dialogic Interaction: Indicated by teachers' open questioning: extended dialogues, and presence of different views.

Open Questions: Questions that have no predefined answer or aim to facilitate students thinking and reasoning.:

Teacher Talk: Teachers' verbal communication: such as lecturing and orchestration of instructional dialogue.

Chapter 7
Different Camera Perspectives, Different Observations?
Pre-Service Science Teachers' Noticing and Perceived Usefulness on Different Classroom Videos

Kennedy Kam Ho Chan
https://orcid.org/0000-0003-1063-2813
The University of Hong Kong, Hong Kong

Steven Ka Kit Yu
The University of Hong Kong, Hong Kong

ABSTRACT

This exploratory study investigated pre-service science teachers' (PSTs') learning in the context of a teacher education activity involving the use of innovative video technology. Specifically, the study investigated the nature of the PSTs' noticing when viewing different videos of their rehearsals of a core practice: eliciting and working with student thinking and PSTs' perceived usefulness of viewing and reflecting on different types of video footage. PSTs' rehearsals were recorded by multiple cameras at various locations, including a video recorder at the rear of the classroom, camera goggles worn by the PST, and a student actor. Findings reveal that the PSTs showed different observation foci, observation specificity, and analytic approach when analyzing different types of footage. Although the PSTs rated viewing all types of footage as useful in enhancing their learning, they gave more favorable ratings on videos filmed from the back of the classroom than the point-of-view footage. The implications of using videos to support PSTs' learning are discussed.

DOI: 10.4018/978-1-6684-6932-3.ch007

BACKGROUND OF THE RESEARCH

The world has become increasingly more complex and uncertain. Teachers are tasked with preparing students to learn to navigate by themselves through a future that is more volatile, uncertain, complex, and ambiguous (Organisation for Economic Cooperation and Development, 2018). Moreover, classrooms have become increasingly more diverse. There is no one-size-fits-all approach in teaching practices. Rather, teachers need to be adaptive and flexible in their approach to meet the needs of diverse learners by engaging in thoughtful and critical reflection on their own practices in their day-to-day practices (Chan & Yung, 2015; Parsons et al., 2018). For this reason, this chapter presents a research study that examined the effectiveness of a new approach to using video technology to enhance teacher learning.

Video technology is widely used to encourage teachers' self-reflection on their practices (Gaudin & Chaliès, 2015; Tripp & Rich, 2012b). Classroom video provides a rich record of classroom events, allowing teachers to revisit their classroom and identify salient events that went unnoticed previously (Sherin, 2004). In this manner, video acts as a mirror to stimulate teachers' reflection (Hollingsworth & Clarke, 2017). Research provides ample evidence that engaging teachers in purposeful viewing and reflection on their video records of instruction can promote their ability to notice salient classroom interactions (Gaudin & Chaliès, 2015; Tripp & Rich, 2012b). These skills represent a process commonly referred to as "teacher noticing," whereby teachers selectively attend to (i.e., what teachers notice) and make sense of particular events in the classroom (i.e., how teachers notice) (Chan et al., 2021a; Sherin, Jacobs, & Philipp, 2011). The ability to notice is essential for teachers to learn from their teaching (e.g., Hiebert, Morris, Berk, & Jansen, 2007; Santagata & Guarino, 2011; Sherin et al., 2011).

Against this background, researchers seeking to document or understand classroom processes have recognized the "keyhole" effect (Sherin & van Es, 2007) of using a single camera. They have advocated the use of multiple cameras placed at different locations to capture multiple points of view of the same classroom to provide a complementary account of classroom processes (Clarke, Mitchell, & Bowman, 2009; Fadde & Rich, 2010). Researchers using videos for professional development of teachers have raised similar concerns. Sherin and Dyer (2017b) note a common myth about video, a medium often considered as objective. They remind us that "whether the video is taken from the front or back of the room, zooms in or is wide angle, influences what can be seen in the video and the sense one gets of the classroom being observed" (p. 49). As such, a video is a record that only reflects a particular point of view of what occurs in a classroom (Sherin & Dyer, 2017a).

Most often, teachers are asked to reflect on video footage filmed from the back of the classroom (Sherin & Dyer, 2017a). Recently, some studies have explored using video footage taken from a point of view other than the back of the classroom for teachers' professional development. Luna and Sherin (2017), for example, engaged elementary science teachers in sharing and discussing footage captured by wearable cameras mounted on teachers' heads to promote the teachers' attention to student ideas in science. Relatively fewer attempts have explored the use of multiple cameras to capture different points of view of the same classroom for enhancing professional learning of teachers. It remains unclear whether viewing and reflecting on videos recorded by cameras at multiple positions would lead teachers to notice different classroom objects and change how they make sense of what they "see" in the footage and which type of footage teachers prefer to watch.

This study addresses this gap by drawing on data from an initial teacher education activity exploring the use of innovative video technology to promote pre-service science teachers' (PSTs) learning from their practices. Specifically, this exploratory study investigated the nature of PSTs' noticing on their rehearsal videos of a core practice recorded by cameras at different positions. Our research was guided by two questions: (1) What is the nature of PSTs' noticing on videos recorded from different positions of the same classroom (i.e., at the rear of the classroom, worn on the head of the PST and that of the student actor)? (2) What are the PSTs perceived usefulness of viewing and reflecting on videos recorded from different positions of the same classroom?

METHODS

Research Approach and Context

The reported study is part of a larger qualitative study aiming to enhance PSTs' learning to enact a core practice: eliciting and working with student thinking (Kloser, 2014). This core practice focuses on drawing out content-specific student thinking and making inferences and responding to it (Chan, 2022; Gotwals & Birmingham, 2016; Windschitl, Thompson, Braaten, & Stroupe, 2012). Nine PSTs (pseudonym used below) voluntarily attended a series of workshops (21 hours) within a science methods course before their student teaching. The workshops introduced the PSTs to the focal core practice using classroom video clips and provided opportunities for the PSTs to rehearse the core practice and reflect on their enactment of the core practices in their rehearsal videos. Further details of the workshop can be found in Chan et al. (2021b). As part of the workshops, the PSTs conducted rehearsals to approximate the core practice (Lampert et al., 2013) and reflected on their rehearsal videos (about 10 to 15 minutes) recorded from different locations, including a video recorder at the back (observer camera-in-the-back (O-CIB) camera), camera goggles worn by the PST (teacher point-of-view (T-POV) camera), and one on a student actor (student point-of-view (S-POV) camera). See Figure 1 for details.

Figure 1. Video camera positions and design of the rehearsal activity

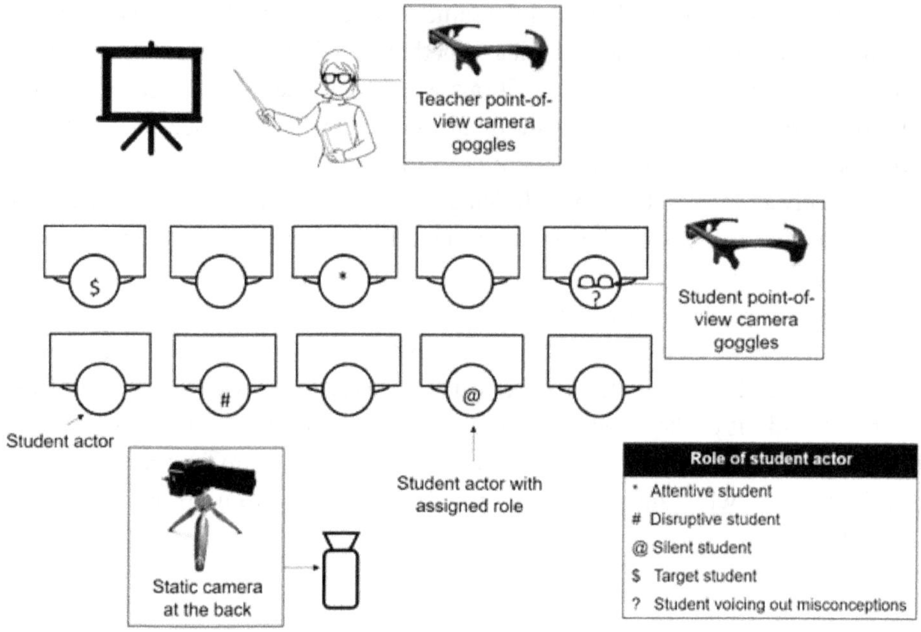

Data Collection

The PSTs completed a *Video Reflection Task* while watching the footage. The task required the PSTs to list the salient aspects in the video in the form of noticing statements (Roller, 2016). The open-ended nature of the prompts allowed us to understand what the PSTs were attuned to notice in different videos. The PSTs first reflected on their CIB-V footage, then T-POV footage, and S-POV footage. They spent about 45 minutes on each video.

To elicit PSTs' views about the usefulness of the different learning activities within the workshops, including the use of different types of video footage, the PSTs completed an anonymous *Written Survey*. In the survey, they rated, on a scale of 1 to 5, the usefulness of the activities (1 = Not at all useful; 2 = Slightly useful; 3 = Somewhat useful; 4 = Very useful and 5 = Extremely useful) in enhancing their learning of the focal core practice. They were also asked to identify the three most useful activities among the 18 activities in the intervention and explain why they felt these activities were useful. Of relevant to this part of the research is their ratings on three items including the usefulness of (1) Viewing and reflecting on your rehearsal video (i.e., observer camera-in-the-back (O-CIB) camera); (2) Viewing and reflecting on your point of view (POV) video footage (i.e., teacher point-of-view (T-POV) camera); and, (3) Viewing and reflecting on student perspective point of view (POV) video footage (i.e., student point-of-view (S-POV) camera).

Data Analysis

For RQ#1, we used codes to characterize the noticing statements in terms of the PSTs' (1) observation focus, (2) observation specificity ((1) and (2) pertain to what the PSTs noticed), and (3) the stance (i.e., how the PSTs noticed) based on the noticing literature (e.g., Chan & Sau, 2021; Lam & Chan, 2020; Tekkumru-Kisa & Stein, 2015; van Es & Sherin, 2008). See Table 1 for details. The frequency of each category is reported in Tables 2 and 3. The first author and a research assistant independently coded each statement. The inter-rater reliability, given by Cohen's kappa, ranged from 0.82 to 1.0. Discrepancies were resolved through discussion.

Table 1. Code list

<table>
<tr><th colspan="2">Observation focus</th><th>Description</th><th>Example</th></tr>
<tr><td colspan="2">Classroom management</td><td>• Comments about the ways the teacher dealt with discipline events, the teacher's procedures for calling on students or attention to different students, the pace and timing of the lesson, and the teachers' physical presence (e.g., patterns of moving around the classroom, tone and volume of voice)</td><td>• I stood at the same position and moved my head for most of the time. (Peter T-POV footage)</td></tr>
<tr><td colspan="2">Teacher pedagogy</td><td colspan="2">• Comments about the actions and strategies used</td></tr>
<tr><td colspan="2">Teacher pedagogy not tied to students</td><td>▪ Teacher's actions/strategies with no explicit relation to students' ideas or actions</td><td>▪ I noticed that I used activity and video to enhance students' engagement and introduce the topic. (Renee, O-CIB footage)</td></tr>
<tr><td colspan="2">Teacher pedagogy tied to students' actions or ideas at a general, non-content-specific level</td><td>▪ Teacher's actions/strategies in some relation to students' ideas or actions but at a general level without having any content specificity</td><td>▪ I addressed the misconception of the students when I found it from what they said. (Joyce, O-CIB footage)</td></tr>
<tr><td colspan="2">Teacher pedagogy in relation to students' actions or ideas at a content-specific level</td><td>▪ Teacher's actions/strategies that got influenced from something specific that student(s) said/did</td><td>▪ I used that [student] question to construct the understanding of the [wall] thickness difference between the atrium and the ventricle. (Barry, O-CIB footage)</td></tr>
<tr><td colspan="2">Student engagement</td><td>• Comments about the level of student participation/students' interest</td><td>• The student was drawing pictures on the textbook. I noticed he was off task, but I didn't know he was drawing cartoons. (Anna, S-POV footage)</td></tr>
<tr><td colspan="2">Student thinking</td><td colspan="2">• Comments about students' cognitive understanding/thinking</td></tr>
<tr><td colspan="2">Student thinking (general)</td><td>▪ Comments about the general understanding or reasoning of students</td><td>▪ I noticed students were able to speak out the correct answer of the question. (Renee, S-POV)</td></tr>
<tr><td colspan="2">Student thinking (specific)</td><td>▪ Comments about specific student ideas or reasoning/ specific ideas students appear to think/understand</td><td>▪ Some students could draw a simple heart shape to represent their understanding, but some could even draw a complex and correct heart structure diagram. (Barry, O-CIB footage)</td></tr>
<tr><th colspan="2">Observation specificity</th><th>Description</th><th>Example</th></tr>
<tr><td rowspan="2">Inclusion of science content</td><td>No content</td><td>• Comments did not mention any science content/ideas</td><td>• I noticed that I did not provide answer to the students. (Renee, O-CIB footage)</td></tr>
<tr><td>With content</td><td>• Comments involved the science content</td><td>• I noticed that I tried to use drawing to elicit students' prior knowledge about the heart structure. (Barry, O-CIB footage)</td></tr>
<tr><td rowspan="2">Description specificity</td><td>General</td><td>• Statements that examined a whole class/group of students or that were broad generalizations</td><td>• I noticed that students felt happy when I picked them to share their ideas. (Barry, S-POV footage)</td></tr>
<tr><td>Specific</td><td>• Statements that were specific to particular events/ideas/individuals/ issues. This may include one student's idea or a particular teacher move</td><td>• I observed that a student was daydreaming, but I didn't pay much attention on him. (Fanny, O-CIB footage)</td></tr>
<tr><th colspan="2">Analytic approach</th><th>Description</th><th>Example</th></tr>
</table>

Continued on following page

Table 1. Continued

Observation focus		Description	Example
Stance: Teacher's analytic approach in commenting on the video/ responses to the video clip	Descriptive	• Refers to statements in which the teacher recounted events that occurred in the video or the teacher made observations about things that student(s) said or did and recounted what was said/done by the student(s)	• Most of the time I was focusing on the whiteboard. (Peter, T-POV) • The silent student was not focusing [on the lesson] and mainly focused on her own work. (Annie, S-POV)
	Evaluative	• Refers to statements in which the teacher commented on what was good or bad or what should have been done differently. • Refers to statements in which the teacher evaluated the accuracy of student ideas	• I noticed that most of the students were not very active in answering questions; the learning atmosphere was not good enough. (Gloria, O-CIB footage) • I need to arrange the question session better by telling them that everyone needs to give an answer before I start asking them questions. (Annie, O-CIB)
	Interpretive	• Refers to statements in which the teacher tried to make inferences about what he/she observed. • Refers to statements in which the teacher examined the meaning of the student ideas, make conjecture about what students understood and speculated why students thought in a particular way	• I noticed my initial explanation of different concepts was not clear. Hence, students only showed evidence of understanding of the concepts after I had rephrased it and explained again. (Tom, O-CIB) • I noticed for most of the questions, only one student answered the question. I could not know if majority of them understood the concept. (Tom, O-CIB)

For RQ#2, we calculated the average ratings of the three items related to viewing and watching different types of footage. We also identified the number of PSTs who chose the activity as one of the three influential activities in their learning of the focal core practice. If the activity was ranked as the most influential activity, it was given a score of 3. The second-ranked activity was given a score of 2, and the third-ranked activity was given a score of 1.

FINDINGS

We organize our findings in relation to the two research questions. We first present findings related to the PSTs' observation foci when watching different footage and then their perceived usefulness of different types of footage.

RQ1: Observation focus when watching different types of footage

Table 2. PSTs' observation focus when reflecting on different types of video footage

		Observer camera-in-the-back footage *(n=42)*	**Teacher point- of-view footage** *(n=21)*	**Student point-of-view footage** *(n=18)*
Focus of observation	Classroom management	9 (21.4)	18 (85.7)	4 (22.2)
	Teacher pedagogy	26 (61.9)	2 (9.5)	0 (0.0)
	Teacher pedagogy not tied to students	*16 (38.1)*	*2 (9.5)*	*0 (0.0)*
	*Teacher pedagogy tied to students' actions or ideas at a general, non-content-specific level**	*8 (19.0)*	*0 (0.0)*	*0 (0.0)*
	Teacher pedagogy in relation to students' actions or ideas at a content-specific level#	*2 (4.8)*	*0 (0.0)*	*0 (0.0)*
	Student engagement	3 (7.1)	1 (4.8)	13 (72.2)
	Student thinking	4 (9.5)	0 (0.0)	1 (5.6)
	Student thinking (General)@	*2 (4.8)*	*0 (0.0)*	*1 (5.6)*
	Student thinking (Specific)+	*2 (4.8)*	*0 (0.0)*	*0 (0.0)*

Notes: The number of idea units is given in the table. The percentages follow in parentheses.

* Refers to teacher's actions in relation to students' actions/ideas without mentioning any science content (Tekkumru-Kisa & Stein, 2015)

Refers to teacher's actions in relation to students' actions/ideas with a consideration of science content (Tekkumru-Kisa & Stein, 2015)

@ Refers to general understanding or reasoning of students in the class (Levin & Richards, 2011)

+ Refers to specific student ideas or reasoning/specific ideas students appear to think/understand (Levin & Richards, 2011)

Table 2 shows the PSTs' *observation focus* when watching different footage. PSTs showed a predilection of focusing on teacher pedagogy when watching the O-CIB footage (26/42). The following are some illustrative statements:

My guiding questions are usually narrow such that students only have a little space to think. (Peter, O-CIB footage)

I noticed that my questions were very academic; it would be better if I could relate [my question] to more everyday examples. (Gloria, O-CIB footage)

Above, Peter identified the weakness of his elicitation questions, whereas Gloria suggested how her questions could have been improved. Notice that both PSTs focused on their own pedagogy with no explicit relation to student ideas/actions "visible" in the video. This type of observation focus was the most common among the statements on O-CIB footage (16/42).

The PSTs' observation focus changed when watching the T-POV clips, as exemplified by the following statements:

When I raised a question, I tended to move towards a particular student. (Gloria, T-POV footage)

I noticed that I liked to pick students who were inattentive to answer my questions. (Barry, T-POV footage)

I focused too much on those disruptive students in the front. (Joyce, T-POV footage)

The PSTs mostly wrote statements about how they moved around the room (e.g., Gloria), called on students (e.g., Barry) or managed students (e.g., Joyce) when reflecting on the T-POV footage. Although the clip provided visual access to "students", the major focus of the statements was not about students. Rather, they were mostly about how the PST *themselves* handled the class (18/21).

We found a significant shift in the PSTs' observation focus when they were watching the S-POV footage. Consider what Joyce and Fanny wrote below:

The student was doing her own stuff and talking to her neighbors, but I did not notice that. (Joyce, S-POV footage)

I noticed that the student just occasionally paid attention to me. Most of the time, this student kept drawing and looking around. (Fanny, S-POV footage)

Interestingly, although the S-POV footage did not capture the images of the student who wore the camera goggles, both PSTs focused on that particular student. Yet, these statements, like most others on S-POV footage, were not about students' cognitive thinking (1/18) but their engagement levels (13/18).

Table 3. PSTs' specificity of observations and analytic approach (i.e., stance) when reflecting on different types of video footage

<table>
<tr><th></th><th></th><th></th><th>Observer camera-in-the-back footage
(n=42)</th><th>Teacher point-of-view footage
(n=21)</th><th>Student point-of-view footage
(n=18)</th></tr>
<tr><td rowspan="4">Specificity of observations</td><td rowspan="2">Inclusion of science content</td><td>No content</td><td>33 (78.6)</td><td>21 (100.0)</td><td>18 (100.0)</td></tr>
<tr><td>With content</td><td>9 (21.4)</td><td>0 (0.0)</td><td>0 (0.0)</td></tr>
<tr><td rowspan="2">Description specificity</td><td>General</td><td>31 (73.8)</td><td>18 (85.7)</td><td>8 (44.4)</td></tr>
<tr><td>Specific</td><td>11 (26.2)</td><td>3 (14.3)</td><td>10 (55.6)</td></tr>
<tr><td colspan="2" rowspan="3">Analytic approach (Stance)</td><td>Descriptive</td><td>26 (61.9)</td><td>19 (90.5)</td><td>17 (94.4)</td></tr>
<tr><td>Evaluative*</td><td>9 (21.4)</td><td>0 (0.0)</td><td>1 (5.6)</td></tr>
<tr><td>Interpretive</td><td>7 (16.7)</td><td>2 (9.5)</td><td>0 (0.0)</td></tr>
</table>

Notes: The number of idea units is given in the table. The percentages follow in parentheses.

* Refers to statements in which the teachers commented on what was good or bad or commented on what should have been done differently. (van Es & Sherin, 2008)

Table 3 displays the PSTs' *observation specificity* and the analytic approach they adopted. It can be seen that all the statements related to the T-POV and S-POV footage did not reference any science. The statements on T-POV footage typically concerned the whole class but not an individual student, as typified by the following example:

I didn't pay attention to all of the students in the class. (Anna, T-POV footage)

The PSTs wrote more specific description on S-POV footage (10/18) than the T-POV footage (3/21). Their statements on S-POV clips often focused on a particular student, usually the one who wore the POV goggles. Annie's statement illustrates this type of observation well:

The silent student seemed to be more interested in the interesting content, which could be some examples she had never known or something out of the (boring) textbook. (Annie, S-POV footage)

While most of the statements about the O-CIB footage were not specific in description (31/42) and devoid of science content (33/42), we did find some statements that were specific in description and included science content, as exemplified below:

I noticed some of my guiding questions had not been explained clearly; therefore students found it hard to answer (e.g., "Which of these two values is large?" should be changed to "You can see that one value is higher, but do you know the actual meaning of a higher value?"). (Tom, O-CIB footage)

One student asked the question, "Why does the heart need so many chambers? Why doesn't it just get one space so that it can receive and pump blood immediately?" (Barry, O-CIB footage)

It is noteworthy that only when watching the O-CIB footage did PSTs write statements that mentioned the science content (9/42) or related to students' content-specific thinking (2/42) (see Barry's statement above and Table 1).

For *analytic approach*, most noticing statements were descriptive in nature (O-CIB-V: 26/42; T-POV: 19/21; S-POV: 17/18). In some occasions, when watching the O-CIB clips, the PSTs interpreted their observations (7/42; see Peter's statement above) or suggested what should have done differently (9/42; see the statements of Gloria and Tom above).

In summary, we found that the observation foci and specificity differed when the PSTs were watching different types of footage. Their analytic stance was also different when viewing different footage. The nature of PSTs' noticing was found to be malleable as a result of the different camera perspectives.

RQ2: Perceived usefulness of the different types of footage

Table 4. PSTs' views of the usefulness of the activities in enhancing their learning of the focal core practice

Activity			Rating of usefulness	
			Total PSTs (n=9) Mean (SD)	**Score* (PST) #**
Viewing and reflecting on one's own rehearsal	(a)	Viewing and reflecting on your rehearsal video (i.e., observer camera-in-the-back (O-CIB) camera footage)	4.2 (0.7)	9 (5)
Viewing and reflecting on POV footage	(b)	Viewing and reflecting on your point of view (POV) video footage	3.8 (0.8)	0 (0)
	(c)	Viewing and reflecting on student perspective point of view (POV) video footage	3.9 (0.8)	0 (0)

* 1 = Not at all useful; 2 = Slightly useful; 3 = Somewhat useful; 4 = Very useful and 5 = Extremely useful

The PSTs chose three activities that they perceived to be most useful to their learning of the focal core practice. The top-ranked activity was given a score of 3, the second-ranked activity was given a score of 2, and the third-ranked activity was given a score of 1. The number in the brackets refers to the number of PSTs who selected this activity as one of the three activities.

Table 4 shows PSTs' rating of the usefulness of the different types of footage in their learning of the focal core practice. Viewing and reflecting on one's own rehearsal footage (i.e., O-CIB footage) was rated more favorably (Rating = 4.2) than viewing and reflecting on two types of POV footage (T-POV = 3.8; S-POV = 3.9). The PSTs' perceived usefulness of watching the O-CIB footage was more apparent when considering their responses on the three most useful activities within the workshop; More than half (i.e., 5 out of 9 PSTs) ranked viewing and reflecting on one's own rehearsal videos as one of the three most useful activities while *none* of them regarded viewing and reflecting on the POV footage as such.

As none of the PSTs chose viewing and reflecting on the two types of POV footage as one of the three most influential activities, we do not have qualitative data about PSTs' views on role of viewing the POV footage in enhancing their learning of the focal core practice. Amongst the five responses that explained why learning from O-CIB camera footage was useful, three of them referred to the affordances of viewing and reflecting on own video in general terms. Below shows one representative example:

After analyzing my previous rehearsal video, I saw many rooms for improvement. Repeatedly watching it reminded me to keep the good things that I did and avoid making the mistakes that I made before. (PST-5, Written Survey)

Above the PST was referring to the affordances of repeated reviewing and reflecting on video of one own rehearsal practices. The video provided opportunities for the PST to step back as observers when re-engaging with particular moments in class (Tripp & Rich, 2012a). In so doing, the PST was able to identify desirable and not so desirable enactment of the focal core practices for future improvement.

Two of the five responses specifically referred to the importance of being able to identify both the teacher instruction and the student responses in the footage (italicized for emphasis below). See the two responses below:

By recognizing my instruction and students' response and reaction, I was able to evaluate the effectiveness of the formative assessment approach I adopted. Then I reflected and analyzed the (students') difficulties and what can be improved. (PST-4, Written Survey))

I rewatched the video and noticed what I did not notice in class. I also noticed that I seldom used students' ideas and targeted their misconceptions. I tried to improve in the second rehearsal (PST-6, Written Survey)

It is worth noting that in both responses, the PSTs mentioned noticing student ideas, such as their difficulties and misconceptions. These data suggest that the O-CIB footage may provide a camera perspective that allowed the PSTs to view the whole classroom scene which might have been easier for them to see *both* the teacher actions and the student responses. These reported benefits were in line with data related to their observation foci (Table 2) which reveals that the PSTs were drawn to teacher pedagogy and/or teacher pedagogy tied to student actions. As it is important for PSTs to reflect on teacher actions in relation to student responses when it comes to learning the core practice of eliciting and working with student thinking, it appears that the O-CIB footage is more useful in this regard than the POV footages.

To summarize, the PSTs perceived viewing and reflecting on O-CIB footage more useful to their learning of the focal core practice than viewing T-POV or S-POV footage. Their reasons for why they learnt so much from the O-CIB footage led us to speculate that this camera perspective directed the focus to both teacher actions and student responses which may hence be regarded as more useful to learning the focal core practice.

DISCUSSION

The purpose of this study was two-fold; to identify, if any, differences in (1) PSTs' observation foci when viewing and reflecting on different types of video footage (i.e., CIB, T-POV and S-POV); (2) perceived usefulness of viewing these three types of videos in their learning of the core practice of eliciting and working with student thinking. Our study revealed some findings that are useful for teacher educators who would like to use innovative video technology, particularly video records captured by multiple cameras of the same classroom, for enhancing professional learning of PSTs.

For RQ#1, our data revealed what the PSTs noticed and how they noticed differed when they were viewing and reflecting on videos filmed from different positions (Table 2, Table 3). Two points are worth further elaboration. First, although the T-POV clips did not provide visual access to the teacher (i.e., the PST who conducted the rehearsal), the PSTs tended to focus on themselves when watching the footage. Likewise, even though the S-POV clips did not capture the images of the student who wore the camera goggles, the PSTs placed their observation focus on that student. It seems that it is not what the PSTs could "see" in the footage, but the first-person perspective of the camera that effectively directed their observation focus. Second, we did not find any statements that referenced science content or specific student ideas when the PSTs were watching the first-person perspective footage (i.e., POV footage). Only when the PSTs were watching footage from a third-person perspective (i.e., O-CIB footage), we found that they noticed specific student ideas. It appears that the PSTs were drawn to the motions/actions captured by the footage, effectively shifting their focus away from the cognitive thinking of the students.

For RQ#2, the PSTs rated viewing and reflecting on CIB footage more favorably (Table 4). For example, while more than half of the PSTs ranked CIB activity as one of the three most influential learning experiences none included watching POV footage. These data revealed that the PSTs perceived viewing and reflecting on the CIB footage more useful than the POV footage. These data are in line with data on their observation foci in different types of footage (Table 2). These data led us to re-examine our assumption about the added value of engaging PSTs in viewing and reflecting on POV footage. At the outset, we thought that by engaging our PSTs in viewing and reflecting on their rehearsals via other classroom perspectives made available in the two POV clips (i.e., first-person teacher's POV and first-person students' POV) would be beneficial to their learning. This is based on our assumption that analyzing the interactions and responses from multiple viewpoints would help the PSTs to reflect on their enactment of the core practice from the perspectives of the teacher and students. The responses from the PSTs led us to wonder whether that was the case. In hindsight, we see that POV footage is an extremely information-dense medium; it does not only capture rich images of the classroom events, but also the motions and actions of the actor. The information-dense footage may easily lead to cognitive overload (Erickson, 2007). Our PSTs' substantive focus on the student engagement but not students' cognitive thinking further alerted us to the difficulty of noticing salient features in POV footage. In other words, the rich information might have side-tracked our PSTs and shifted their attention to the motions/actions of the actors away from the focus of the interactions between the multiple actors in the classrooms. We might have under-estimated the difficulties of adopting the perspective of the respective actor simply by watching a POV clip and did not provide sufficient support and scaffolds to guide our PSTs to notice the interactions in the POV footage that were salient for analysis. This difficulty led us to question whether a static camera placed at the back of the classroom could capture sufficient detail and render the POV footage unnecessary. Given the dearth of research on the use of POV video clips to enhance teacher learning (see a prominent exception, Luna & Sherin (2017)), this exploratory study provides teacher educators with a unique opportunity to think critically about the unique affordances and limitations of POV video and how it can complement the use of video taken with a static camera in the back of the classroom in enhancing teacher learning. We are of the view that by opening up various windows to the same classroom scene via the multiple cameras, we can allow the PSTs to analyze different features of the classroom. Perhaps, what is more important is to help the PSTs to integrate and connect what they can notice in different video records to form a holistic interpretation of what transpires in the classroom interactions in terms of their enactment of core practices with a focus on the effect of teacher moves on student cognitive thinking. Such activity might help PSTs look beyond the trivial information and form a more holistic interpretation of what happens in the complex classroom scene.

Implications and Limitations

Our data contributes to a better understanding of how different camera perspectives may privilege certain type of PSTs' noticing. For example, when watching S-POV clips, the PSTs shifted their observation focus to the student who wore the camera goggles. Our findings also contribute to a better understanding of the affordances and limitations of different types of video in the context of promoting PSTs' learning through video-based reflection on their rehearsal experience. For instance, although the PSTs shifted their observation focus to the students when viewing the S-POV clips, their attention was mostly drawn to the actions/motions of the students rather than their thinking. These findings and insights contribute to the growing interest of using video footage captured from different locations of the classroom for professional development of teachers (e.g., van Es, Stockero, Sherin, Van Zoest, & Dyer, 2015). Our data also caution that teacher educators need to carefully consider the unique affordances and added value of using POV footage in teacher education activities and to think of how to help PSTs integrate and connect what they see in different cameras to form a more high-level and holistic interpretation of the classroom scene when using multiple cameras.

In closing, we acknowledge the limitations of our study, including the small sample size which makes our findings not generalizable. We also concede that the nature of the PSTs' noticing can be influenced by a number of factors (e.g., the noticing ability of the PSTs, the sequence of watching different types of footage). Although it is clear from our data that the camera perspective is one of the contributing factors to the different nature of PSTs' noticing we observed, future study with larger sample and in diverse contexts can verify our tentative findings. Moreover, as we did not collect qualitative data on the unique affordances of viewing and reflecting on different types of POV footage. Future studies should also further delineate the unique affordances of these two types of POV footage.

ACKNOWLEDGEMENT

We acknowledge the funding supports by the Hong Kong Research Grants Council, General Research Fund Grant/Award Number: 17611620.

REFERENCES

Chan, K. K. H. (2022). Eliciting and working with student thinking: Preservice science teachers' enactment of core practices when orchestrating collaborative group work. *Journal of Research in Science Teaching*, *60*(5), 1014–1052. doi:10.1002/tea.21823

Chan, K. K. H., Xu, L., Cooper, R., Berry, A., & van Driel, J. H. (2021a). Teacher noticing in science education: Do you see what I see? *Studies in Science Education*, *57*(1), 1–44. doi:10.1080/03057267.2020.1755803

Chan, K. K. H., & Yau, K. W. (2021). Using video-based interviews to investigate pre-service secondary science teachers' situation-specific skills for informal formative assessment. *International Journal of Science and Mathematics Education*, *19*(2), 289–311. doi:10.100710763-020-10056-y

Chan, K. K. H., Yu, S. K. K., & Sin, R. K. H. (2021b). Using student actors and video-mediated reflection to promote pre-service teachers' ability to enact responsive teaching. *Innovations in Science Teacher Education*. *6*(1). https://innovations.theaste.org/using-student-actors-and-video-mediated-reflection-to-promote-preservice-teachers-ability-to-enact-responsive-teaching/

Chan, K. K. H., & Yung, B. H. W. (2015). On-site pedagogical content knowledge development. *International Journal of Science Education*, *37*(8), 1246–1278. doi: 10.1080/09500693.2015.1033777

Clarke, D., Mitchell, C., & Bowman, P. (2009). Optimising the use of available technology to support international collaborative research in mathematics classrooms. In Janik. T. & T. Seidel (Eds.), The Power of Video Studies in Investigating Teaching and Learning in the Classroom (pp. 39-60). New York: Waxmann.

Erickson, F. (2007). Ways of seeing video: Toward a phenomenology of viewing minimally edited footage. In R. Goldman, R. Pea, B. Barron, & S. J. Derry (Eds.), *Video Research in the Learning Sciences* (pp. 145–158). Lawrence Erlbaum.

Fadde, P., & Rich, P. (2010). Guerrilla video: A new protocol for producing classroom video. *Educational Technology*, *50*(1), 4–8.

Gaudin, C., & Chaliès, S. (2015). Video viewing in teacher education and professional development: A literature review. *Educational Research Review*, *16*, 41–67. doi:10.1016/j.edurev.2015.06.001

Gotwals, A. W., & Birmingham, D. (2016). Eliciting, identifying, interpreting, and responding to students' ideas: Teacher candidates' growth in formative assessment practices. *Research in Science Education, 46*(3), 365–388. doi:10.100711165-015-9461-2

Hiebert, J., Morris, A. K., Berk, D., & Jansen, A. (2007). Preparing teachers to learn from teaching. *Journal of Teacher Education, 58*(1), 47–61. doi:10.1177/0022487106295726

Hollingsworth, H., & Clarke, D. (2017). Video as a tool for focusing teacher self-reflection: Supporting and provoking teacher learning. *Journal of Mathematics Teacher Education, 20*(5), 457–475. doi:10.100710857-017-9380-4

Kloser, M. (2014). Identifying a core set of science teaching practices: A Delphi expert panel approach. *Journal of Research in Science Teaching, 51*(9), 1185–1217. doi:10.1002/tea.21171

Lam, D. S. H., & Chan, K. K. H. (2020). Characterising pre-service secondary science teachers' noticing of different forms of evidence of student thinking. *International Journal of Science Education, 42*(4), 576–597. doi:10.1080/09500693.2020.1717672

Lampert, M., Franke, M. L., Kazemi, E., Ghousseini, H., Turrou, A. C., Beasley, H., & Crowe, K. (2013). Keeping it complex: Using rehearsals to support novice teacher learning of ambitious teaching. *64*(3), 226-243. https://journals.sagepub.com/doi/abs/10.1177/0022487112473837

Levin, D. M., & Richards, J. (2011). Learning to attend to the substance of students' thinking in science. *Science Educator, 20*(2), 1.

Luna, M. J., & Sherin, M. G. (2017). Using a video club design to promote teacher attention to students' ideas in science. *Teaching and Teacher Education, 66*, 282–294. doi:10.1016/j.tate.2017.04.019

Organisation for Economic Cooperation and Development. (2018). *The future of education and skills: Education 2030*. OECD.

Parsons, S. A., Vaughn, M., Scales, R. Q., Gallagher, M. A., Parsons, A. W., Davis, S. G., Pierczynski, M., & Allen, M. (2018). Teachers' instructional adaptations: A research synthesis. *Review of Educational Research, 88*(2), 205–242. doi:10.3102/0034654317743198

Roller, S. A. (2016). What they notice in video: A study of prospective secondary mathematics teachers learning to teach. *Journal of Mathematics Teacher Education, 19*(5), 477–498. doi:10.100710857-015-9307-x

Santagata, R., & Guarino, J. (2011). Using video to teach future teachers to learn from teaching. *ZDM Mathematics Education, 43*(1), 133–145. doi:10.100711858-010-0292-3

Sherin, M. G. (2004). New perspectives on the role of video in teacher education. In J. E. Brophy (Ed.), *Using video in teacher education* (pp. 1–27). JAI.

Sherin, M. G., & Dyer, E. B. (2017a). Mathematics teachers' self-captured video and opportunities for learning. *Journal of Mathematics Teacher Education, 20*(5), 477–495. doi:10.100710857-017-9383-1

Sherin, M. G., & Dyer, E. B. (2017b). *Teacher self-captured video: Learning to see.* SAGE. https://journals.sagepub.com/doi/abs/10.1177/0031721717702632

Sherin, M. G., Jacobs, V., & Philipp, R. (2011). *Mathematics teacher noticing: Seeing through teachers' eyes.* Routledge. doi:10.4324/9780203832714

Sherin, M. G., & van Es, E. A. (2007). *Using video to document changes in teachers' professional vision.* Paper presented at the Annual meeting of the American Educational Research Association, Chicago, IL.

Tekkumru-Kisa, M., & Stein, M. K. (2015). Learning to see teaching in new ways. *American Educational Research Journal, 52*(1), 105–136. doi:10.3102/0002831214549452

Tripp, T. R., & Rich, P. J. (2012a). The influence of video analysis on the process of teacher change. *Teaching and Teacher Education, 28*(5), 728–739. doi:10.1016/j.tate.2012.01.011

Tripp, T. R., & Rich, P. J. (2012b). Using video to analyze one's own teaching. *British Journal of Educational Technology, 43*(4), 678–704. doi:10.1111/j.1467-8535.2011.01234.x

van Es, E. A., & Sherin, M. G. (2008). Mathematics teachers' "learning to notice" in the context of a video club. *Teaching and Teacher Education, 24*(2), 244–276. doi:10.1016/j.tate.2006.11.005

van Es, E. A., Stockero, S. L., Sherin, M. G., Van Zoest, L. R., & Dyer, E. B. (2015). Making the most of teacher self-captured video. *Mathematics Teacher Educator, 4*(1), 6-19. https://www.jstor.org/stable/10.5951/mathteaceduc.4.1.0006

Windschitl, M., Thompson, J., Braaten, M., & Stroupe, D. (2012). Proposing a core set of instructional practices and tools for teachers of science. *Science Education, 96*(5), 878–903. doi:10.1002ce.21027

Chapter 8
Preparing Science Teachers for Tomorrow With Transversal Competencies

Rachel S. Sheffield
https://orcid.org/0000-0003-2385-6381
Curtin University, Australia

Rekha Koul
https://orcid.org/0000-0002-7356-0370
Curtin University, Australia

ABSTRACT

The education system necessitates preparing today's children for the needs of tomorrow, and more so in our changing society educated by a schooling system that is agile and flexible in preparing students for jobs yet to be invented. Education systems may no longer be required to teach students content, and instead need to focus on giving students skills to curate information and demonstrate their capacity to be critical and creative thinkers, problem solvers, communicators, and collaborators. Moreover, these skills will be predominantly useful for knowledge acquisition and the creation of science subjects. Are we preparing students for this new world of ever-increasing demands for knowledge? Are we creating thinkers and active learners? This chapter determines how we can create learners who can demonstrate these transversal competencies and substantiates our claims based on research findings where authors searched beyond 'knowledge concepts' and sought to create more agile and flexible science learners who are prepared for some of the challenges posed by the 4th Industrial revolution.

DOI: 10.4018/978-1-6684-6932-3.ch008

INTRODUCTION

The gap between the industry's needs and the scope and quality of competencies offered generally by education is revealing the increasing need to find new educational models to face the new era of the smart digital connected world. (Tan et al., 2018)

The education system necessitates preparing today's children for the needs of tomorrow and more so in our changing society. Children need to be educated by a schooling system that is agile, forward looking, and flexible - preparing students for jobs that are yet to be invented. As society moves through the 4th Industrial Revolution (industry 4.0), education systems are experiencing times of incredible shift in classroom practices developed during the previous three industrial revolutions, and this needs to be given due consideration. In the 20th Century, educators had a sense of power and control, and a notion that students could learn all that they needed from schools and from educators. It was believed that schools contained all the required knowledge. In the last three decades, it has become clear that with the rise of Industry 4.0 there is a continued exponential increase in the amount of knowledge produced in our information society. Educators seek to provide students with the knowledge, which is getting bigger by the day. Knowledge can be quickly and easily accessed using Information Communication Technologies (ICT) tools, available in many everyday use items such as smartphones.

Since the 16th century when the first free public schools were created in the US and UK, schools have been set up alongside industry and industry has informed school pedagogical practices (Robinson, 2010). School systems were based on the traditional view of academic ability at the time that people were either academic (versed in the classics and deductive reasoning) or non-academic (focused on developing required industrial skills). Even in current schools, there are bells to signify lesson time and little choice about what students learn. In science and other secondary subjects, learning is siloed, and students move through their school years based on their age and not their level of development or interest (Robinson, 2010). Makrides (2019) reported that the education system was unable to keep ahead of societal change, highlighting that an estimated 65% of students entering primary schools will work in new jobs that are yet to be created and that 50% of the technical knowledge studied by an undergraduate student will be obsolete by the time they graduate (McKinsey Global Institute, 2017). The Organisation for Economic Co-operation and Development (OECD) (2020) acknowledges that our complex and rapidly evolving world requires a reimagining of education content and delivery, including the reorganisation of formal and informal learning environments and a focus on lifelong learning.

One of the contemporary world's most critical issues is quality education for all children. Teachers can best address this problem; however, they face a range of significant challenges seeking to impart motivation, curiosity, knowledge, and skills to a new generation of students. Schools, classrooms, and teachers are no longer the singular sources of knowledge, and as content knowledge continues to exponentially rise, particularly in science, there is a crisis when teaching content is considered the key to student success. As teacher educators, it is our responsibility to support in-service and pre-service teachers (PSTs) to ensure high-quality learning is available in every classroom worldwide.

We need to think differently about our skills, how we couple with new technologies, and most importantly, how we keep up. Our humanity, integrity and creativity are key (PWC Australia, 2020).

Industrial Revolution 4.0

The First Industrial Revolution (IR), defined by steam power including the steam engine, enabled the large-scale manufacturing of goods and products. This focused on centralising production, which moved from villages to cities in the 1780's. From the late 1800s to the early 1900s, the Second Industrial Revolution, defined by the invention of electricity, enabled mass production, the invention of the internal combustion engine, and the production of steel and eventually petroleum. From the 1950s, the Third Industrial Revolution focused on computers, robots, transistors, and microprocessors, with the Internet creating opportunities for mass communications and globalisation. The World Economic Forum (WEF) (2016) has defined the Fourth Industrial Revolution, Industry 4.0 or 4IR (used interchangeably), as a merging of human and machine, and includes artificial intelligence, genome editing, biometrics, renewable energy, 3D printing, and the Internet of Things.

Navigated wisely, the WEF (2016) sees 4IR as bringing smart cities that have the capacity to reduce poverty, enhance standards of living, and improve human health. Smart cities incorporate sustainable energy sources and provide environmental protection, and are defined by more inclusive government processes, social cohesiveness and collaboration. Boston Consulting Group (2022) analysed 15 million job postings and identified that there is a dramatic change in the demand from 41R employers for transversal or transferable skills such as communication, collaboration, critical thinking, and creativity. Deloitte (2017) in their report *Soft Skills for Business Success*, determined that in the future two-thirds of all jobs will be transversal competencies intensive, focusing on skills rather than knowledge.

In line with the Industrial Revolutions, the corresponding education revolutions seek to help learners meet the needs and employment opportunities that the IR provides. Alongside 4IR is the evolving education revolutions travelling more slowly through a series of similar paradigm shifts. As education is evolving, the following table highlights some of the key features of each education version.

Table 1 focuses on the key elements of learner, curriculum, pedagogy, content, assessment, and purpose. Comparisons can be made in each iteration as to their apparent importance. There is a movement from an instructivist to a constructivist approach, where instead of receiving, responding, and regurgitating information, students are connecting, collecting, and curating information to create new knowledge. While the adoption of a constructivist approach has been promoted in the literature for many decades, in schools there seems to be very little real change. There is still evidence of school systems embracing Education 1.0 and Education 2.0, especially in developing countries. Many schools are not equipped to engage students around the 21st century skills or competencies, or to ensure they are agile and responsive to societal and technological change.

Table 1. The evolution of education based on education evolution chart

Attributes	Education 1.0	Education 2.0	Education 3.0	Education 4.0
Model	Download education	Open access education	Knowledge producing education	Innovation producing education
Curriculum	Fixed knowledge base Exams	Variety of knowledge Exams	Knowledge and skill-based	Fluid learning skills and driven by innovation and learners' needs
Approach and Focus	Instructivist (3Rs) Respond, receive and regurgitate	Constructivist 4Cs Communicate, contribute, collaborate and co-creating	Connectivist 3Cs Connecting, collecting, and curating	Adaptive learning driven by AI or machine learning process based on real-time student profiles, built through selective individual and team-driven embodiments in practice through innovation
Educator Role and Actions	Knowledge source	Facilitator teams with students	Learning designer supporting collaborative knowledge creation	Supported by an AI learning portal (super agile and responsive)
Student actions	Passive learners	Active learners	Co-developers and co-researchers Authors, drivers, and assessors of learning experiences	Self-governed autonomous Counsellors and AI co-develop education plans which constantly evolve The input of learning is the major source of technological evolution
Technology use	Internet age starting	E-learning and collaboration Open source starting	Low-cost digital, mobile, Web-driven technologies Use purposely for the selective production of knowledge	Personalised intelligent models, Internet of Things Web-driven e-learning
Space	Rows of desks and chairs facing the teacher's desk in front	Groups of desks and chairs seeing the teacher in front	Groups of seats but no teacher at the front	No teacher's desk in the room. Students can be anywhere and there is no formal seating

While Education 4.0 is still not at the forefront of educational dialogue and implementation, it is gaining traction as a topic for robust discussion. In summary, the key aspects of Education 4.0 are

- Learners as agents of their own learning - driving learning as partners/ collaborators.
- Seamless learning (not bound by weeks, semesters, or locations) that is a personalised pathway for the learner.
- Multiple learning providers (dependent on the needs of the learner).
- Accumulation of learning credentials to create a learner profile that is constantly updated as lifelong learning continues post-formal schooling.
- Constructive digitalisation; and
- Convergence of AI and learning experiences (Immersive technologies).

STEM/Science Education and Education 4.0

STEM has a number of definitions that permeate the learning landscape this includes any learning that encompasses science, or technology, or engineering or mathematics (S.T.E.M) to a much more focused integrated approach that includes science and technology, and engineering and mathematics (STEM or iSTEM) or partial integration. Integrated STEM education is an approach to teaching and learning that combines two or more STEM subjects (science, technology, engineering, and mathematics) based on connections between the subjects and real-world problems. It can involve engineering design based pedagogical approaches that promote integrative STEM thinking and skills. The integration takes place in ways such that knowledge and process of the specific STEM disciplines are considered simultaneously without regard to the discipline, but rather in the context of a problem, project, or task (Cunningham, & Carlsen, 2014).

When is Education 4.0 translated into a science or STEM classroom context there are two distinct aspects including increased choice of content in integrated STEM content with an increased focus on skills?

Increased Choice

- Science learners can choose the science issue or idea they focus on, including how they study, where they study (with industry), and how they choose to be assessed.
- Science learners can mix and match their learning through multiple learning providers (dependent on their learning needs) and build their learning credentials.

Skills Trend

- STEM learners' focus is on developing key skills including 21st Century or transversal competencies within the context of science.
- STEM learners will be digitally competent and expected to work in the convergence of AI and pedagogy in an 'omnichannel' learning environment.

Currently, there is an impasse in all school education but in particular *science/ STEM education*. Exams that are set at the end of schools such as O levels and A levels in the UK, India, or ATAR in Australia, drive what is taught in secondary classrooms where science is separately siloed from mathematics and technology. These exams, which focus on students' understanding of key concepts, drive the secondary curriculum courses, and their aims then drive the learning in the pre-service teacher education courses at university. Whilst there are skills such as critical and creative thinking and collaboration in the Australian Curriculum, these competencies sit outside subjects and outside of reportable assessment and are thus largely ignored. The focus, therefore, is on content knowledge, and in STEM/ science, this is increasing in amount and in complexity. Student learning is regulated and bound by examinations that reward remembering and regurgitation and have few skills embedded, resulting in classrooms that are structured and regimented. Removal of current exams and standardised testing comparisons either internally or externally is important to encourage personalised learning, remove the siloed traditional approach, and increase student agency.

Examples in University Context

A paradigm shift of educational systems is required - one that frees up the regimented timetables and encourages students to collect in capacity and ability rather than by age. The examples below have been implemented as units into an Australian University School of Education program. In these units we explore

1) how a project-based approach was enacted and supported PSTs student skills acquisition through a unit focused on Makerspace (iSTEM Makerspace)
2) how a problem-based approach was enacted and supported PSTs student skills acquisition through a unit focused on the context of UN Sustainable Development Goals and design thinking called Social Issues (iSTEM Social Issues).

At present, these examples of pre-service teacher (PST) education units that are designed to promote fluid learning skills and that are driven by innovation and learners' needs can only be found in the fringes of university teacher education courses. High levels of accreditation in teacher education courses prevent more innovative and exciting approaches to be embedded in core units. Changes in the current school systems would need to be implemented, and these would need to reflect changes in teacher education courses. These changes will lead to acceptance of the suite of iSTEM units.

The units *iSTEM Makerspace* and *iSTEM Social Issues* are future-oriented units seeking to address the critical area of STEM (Science, Technology, Engineering, Mathematics) education, which is an established international focus, and necessary for the successful creation of a modern workforce equipped to tackle unknown working conditions (Koul et al., 2017; Timms et al., 2018). STEM education develops essential skills required by our near-future society, such as analytical thinking and innovation; complex problem-solving; creativity, originality, and analysis; and leadership and social influence (Rand Corporation, 2012; WEF, 2020).

The iSTEM units were designed to explore learning at the edges of the current PST-accredited courses, where the learning seeks to encourage PSTs to develop their agency when designing assessments and creating experiences for future students (Sanders, 2015). The units seek to engage young female PSTs and encourage and support women in STEM education and careers, as research clearly indicates that enrolments from females are severely lacking in STEM courses (Knopke, 2019). The trend of underrepresentation of women within the fields of STEM is a long-term issue (Broadley, 2015; Petherick et al., 2017; Timms et al., 2018), which means those female role models are not available to inspire young girls to pursue these disciplines. A female's motivation to pursue STEM education is determined by the experiences that they received prior to, and during, their lower schooling years (Timms et al., 2018), highlighting the vital work that the PST units are doing to address this critical educational issue. Whilst these units are open to all Education students, as 87% of all participants in the primary PST program are female and 93% of the early childhood program are also female, this then contributes to the majority of the PSTs studying these units.

The first unit focused on Makerspaces and the underpinning pedagogy of the Makerspace that includes the pedagogical approaches of Piaget (1936) on 'constructivism' and Papert on 'constructionism' (1980), and 'play' theories that have been applied to learning pedagogies since the early 1950s. This unit included working in a primary school and PSTs had an opportunity to see their preparation in action.

The second unit (*iSTEM Social Issues)* focused on Problem-Based-Learning (PBL) exploring one of the models of PBL, in this case the Hasso Plattner Design thinking model (Stanford Design School, 2014), in the context of the United Nations Sustainable Development Goals (SDGs) which incorporate transversal competencies

(United Nations Educational, Scientific and Cultural Organization [UNESCO], 2015). PSTs initially completed the inquiry as a student, following the design steps to develop a solution to their identified problem as their first assessment, and then they planned as PSTs about how design thinking/PBL models could be incorporated into the curriculum.

Both units supported the development of transversal competencies or transversal skills including critical and creative thinking, collaboration, and communication skills.

iSTEM Makerspace

The term *Makerspace* has only become popular since 2016 and after the advent of Hackerspace© to an actual physical place termed a 'Place for Making' or 'Makerspace' (Smith, et al., 2013). The idea of engaging with the environment and learning by 'doing'; however, was actually proposed by Dewey (1902) and then again, with the work of Piaget (1936) on constructionism and Papert (1980) on constructionism, predating the new terminology. Unlike other concepts, which have a clear history outlining the development, Makerspaces have grown organically, and there are a number of ways of defining them. In the *iSTEM Makerspace* unit, the focus is on *Making as the approach* where the students create the same object and are encouraged and scaffolded in their learning (see Table 2). This approach enables the teacher to control the object to be generated but students are required to demonstrate their competency when they create it. For example, the creation of a Wigglebot (Image 4) requires an understanding of the content knowledge of energy transfer and electrical circuits, and skills that include critical thinking and possibly communication, collaboration, and perseverance. The use of skills depends on how the task is evaluated and how challenging the task is to the participating child.

Figure 1. Expansion of types of makerspaces
(Koul et al. 2021, adapted from Vuorikari, et al., 2019)

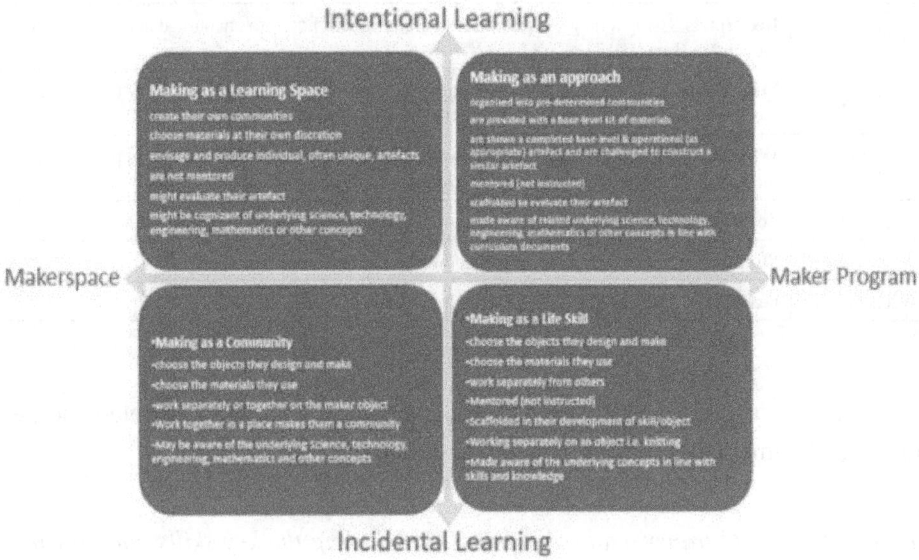

iSTEM Makerspace Context

This unit was run in a primary school with Year 5/6 students.

The PSTs came to their university class one week,

- made the objects,
- designed questions to ask, and
- considered how the creations linked to the STEM curriculum including the Transversal Competencies (TVCs) / General Capabilities as well as concepts.

In the next week, PSTs

- created the objects with students in a class at a local school, and
- used the questions to guide the learning.

In developing the unit, the PSTs reviewed the experiences in class with the teacher educator the following week and engaged in conversations around the skills that had been identified using the TVCs and General Capabilities from the Australian Curriculum and the content knowledge.

Table 2. Description of the makerspace unit

Week	Maker Activities
Week 1 /2	**Bee Hotels.** Using a range of materials to create a bee 'hotel' to encourage native bees into the garden (Image 1).
Week 3/4	**Wigglebots.** Students use their knowledge of electricity and changes in energy to create a moving wobbling bot with a motor, batteries, paper cup and pen legs (Image 4)
Week 5/6	**What can you make out of a 1.25 L bottle?** Student ideated as many ideas as possible to then create their best idea and make it from a 1.25 L bottle (Image 2)
Week 7/8	**Terrarium**. Creating a living microcosm inside a glass bottle with plants and soil.
Week 9/10	**Stomp rocket.** Designing the best air powered rocket from paper (Image 3)
Week 11/12	**Grabber**. Designing the most effective tool to move bottles (Image 5)

In Table 3 examples are provided about how students demonstrated the key transversal competencies in each maker activity.

Table 3. UNESCO transversal competencies including the key skills and how they are represented in a makerspace unit

UNESCO TVC including key skills and competencies	An Example seen in a *Makerspace approach* with students in a primary class.
Critical Innovative. **thinking creatively,** entrepreneurship, resourcefulness, application skills, reflective thinking, decision-making	When generating all the ideas about what they could make with a bottle students, week 5/6 demonstrated their creativity (Image 2).
Interpersonal Skills. communication, leadership, organisational, **teamwork, collaboration, initiative**, collegiality empathy, compassion	Students were required to work in teams to create the **Grabber** (Image 5) therefore demonstrating collaboration.
Intrapersonal Skills. self-discipline, **independent** learning, flexibility, adaptability, self-awareness, **perseverance,** self-motivation, compassion, integrity, risk-taking, self-respect, sense of belonging	When creating the **Wigglebot** students were required to demonstrate perseverance and flexibility to create the circuit (Image 4).
Global Citizenship. awareness, tolerance, openness, respect for diversity, intercultural understanding conflict resolution civic / political participation, **respect for the environment,** national identity	When creating **Bee Hotels**, students were encouraged by PSTs to design a space for bees based on where it will be positioned in the school grounds or at their home. Students were able to choose the materials including branches, wood, tin can and cardboard and consider needs of bees (Image 1).
Media and Information Literacy. accessing information, **locating information, communicating ideas,** analysing information evaluating information and media content	When creating **Bee Hotels** and **Terrariums**, students were required to search for data on the needs of Australian native bees and details on creating terrarium. Students also presented their data on native bees to their PST buddies through a poster.

Not all the identified Transversal competencies were addressed here as **Global Citizenship** e.g. respect for diversity, intercultural understanding conflict resolution civic / political participation, respect for the environment, national identity was not included.

Koul et al. (2021) used the OCED (2020) scenarios that represented Education 1.0 – 4.0, to align the Makerspace approach to one of the scenarios and they determined that the key aspects of the Makerspace aligned best to Education 4.0.

The OCED (2020) predicts a number of possible scenarios for the future of schooling, these four different models suggest.

1. Extended schooling where formal education expands to include international collaboration and individualised learning. The structures and processes of schooling remain as practised in the past.
2. Education is outsourced and privatised with competing providers. This leads to the diversification of opportunities enabled by technology as a key driver for change.
3. Schools become the learning hubs with a connection to the community through face-to-face and online methods. Creating flexible arrangements in schools would support greater personalisation and community involvement. In this scenario, teachers are considered 'knowledge nodes' within wider networks of flexible expertise.
4. Education can be anywhere and anytime with few distinctions between formal and informal education. This scenario is extreme and could include schools being dismantled and teachers becoming 'mobile and active commodities.

While all these scenarios seem far-fetched at present, there are elements of each scenario that are currently in practice.

Figure 2. Features in four phases of education
(Adopted from Koul et. al., 2021)

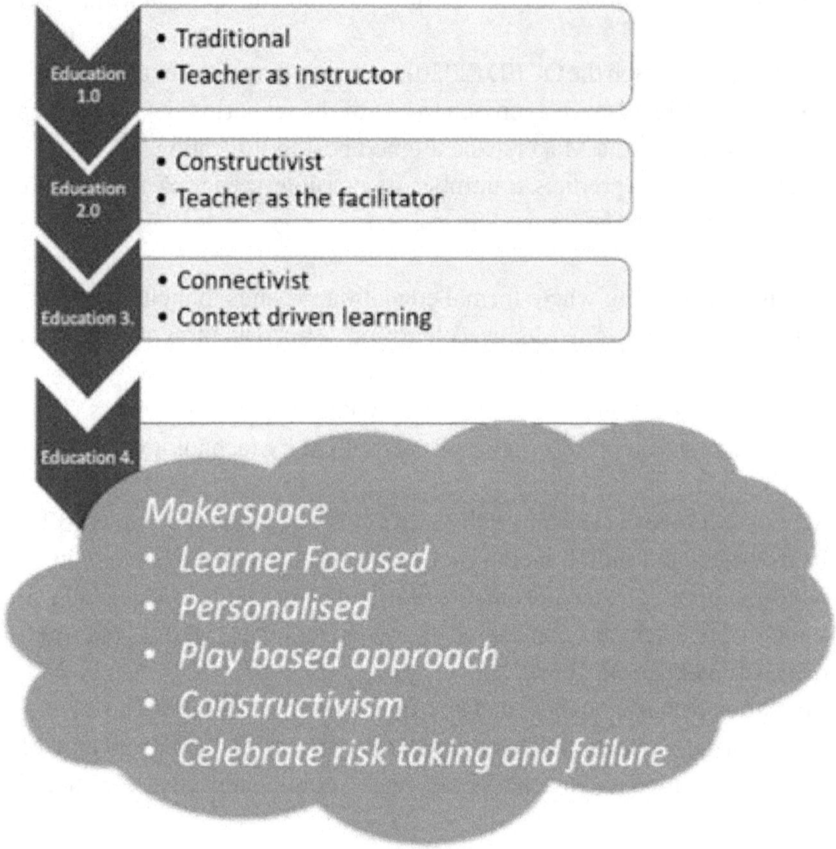

Makerspaces meet every aspect of Education of the Future (4.0) as they are excellent spaces where students can explore topics including those that are based around Science, Technology, Engineering and Mathematics (STEM) areas. Whilst we have not reached this stage in schools yet there is vision that teachers can develop Makerspaces to be utopic places where students can play, create, and think and be part of an engaging and evolving learning ecosystem which fosters the development of key transversal competencies.

iSTEM Social Issues Unit

Design thinking was implemented into the *iSTEM Social Issues* unit, which was developed around the UN SDGs United Nations Sustainable Development Goals.

Problem-based learning and/or design thinking is an approach **or** process used to create *something* to solve a problem (Lewis et al. 1998; Wrigley & Straker, 2017). The *something* can be an idea or a tangible/physical artefact, which solves or addresses the problem that was articulated. Larmer et al. (2015) defined the characteristics of [PBL] as involving students applying content knowledge and skills to collaboratively develop solutions to authentic real-world problems. PBL is multifaceted and complex, requiring more than one lesson to solve the problem.

iSTEM Social Issues Context

Problem based learning can have a number of steps and there are many models. In this situation the unit was developed using the Hasso Plattner (Stanford Design School, 2014) design-thinking model and also the 6D Model which was based on the Hasso Plattner model and developed by the design school of the Hasso-Plattner-Institute at University of Potsdam in Germany (Tschimmel, 2012). We use the terms problem-based learning and design thinking interchangeably here, as in the area of Design, this is *Design Thinking*, whereas in education it can be classified as Problem-Based learning. In this instance the unit incorporated the Hasso Plattner *Design Thinking* model, but when the primary student aspect of the website was created (Image 7) the 6D Model was used; Define, Discover, Dream, Develop, Deliver, Debrief. This model was similar to the Hasso-Plattner model, but the language was considered easier for the primary students to comprehend.

This unit was run in two sections. For the first part, the PST students were required to complete their own DT/PBL based on a topic/issue/problem that they identified in their everyday life. This could include where they worked or in their local community, and needed to be small, researchable, and focused. For example, homelessness in Western Australia was too broad, but reviewing the use and recycling of plastic hangers in a single store of a larger franchise was considered a researchable issue to explore. The PSTs were required to articulate the issue and then to gather data from all stakeholders. For example, with the plastic hangers, the student collected data over four weeks to determine the numbers of hangers used and then created an approximate number used per year. The student then interviewed employees and employers to determine more details about the problem, and how the issue can be resolved. The PSTs created a website to capture each stage of the Hasso Plattner model (Figure 3), based on a website created by the teacher educator at https://istempbl.weebly.com/ (Image 7, Image 8).

Figure 3. Hasso Plattner
(Stanford Design School, 2014)

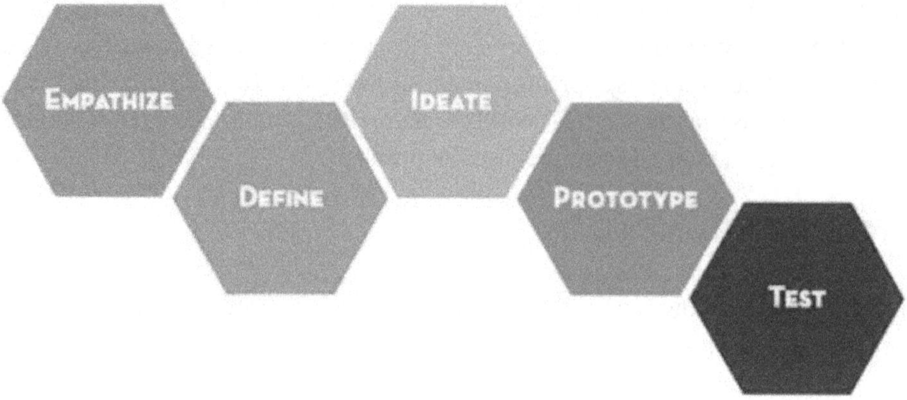

Table 4. Description of the design thinking model stages

Stage of the Model	
Empathize	The *empathize* stage focuses on understanding the values of others and collecting data in the form of surveys, interviews, etc. Other data forms can include environmental scans, quantitative data collection such as a count, or using historical or secondary data about the context of the problem being studied.
Define	The *define* stage explicitly expresses the problem that one strives to address. This is clearer than at the start where the problem may be hazier and less clear.
Ideate	The *ideate* stage is where a range of ideas and concepts are ideated, from blue sky to more mundane ideas. The aim is to create many ideas and then one idea is chosen.
Prototype	The *prototype* stage is when one gets the ideas conceptualised and into the world. It is an example object or artefact that demonstrates the same features and qualities as the final product or solution. This is an experimental phase, and the aim is to identify the best possible solution for each of the problems identified during the first three stages.
Test	The *Test* stage is the chance to gather user feedback, refine solutions, and continue to learn about one's users.
Storytelling	Where the designer gets to share and create a narrative around their design and subsequent solution

Table 5 provides examples of how problem-based learning can support the development of transversal competencies, as demonstrated by the pre-service teachers during the unit.

Table 5. UNESCO (2015) transversal competencies including the key skills and how they are represented in a PBL unit

UNESCO TVC including key skills and competencies	PBL
Critical Innovative and Thinking creatively entrepreneurship, resourcefulness, application skills, **reflective thinking**, decision-making	Ideation of new ideas that encourage students to be creative.
Interpersonal Skills communication, leadership, organisational, **teamwork, collaboration, initiative**, collegiality empathy, compassion	Teams of students working on the project together focusing on supporting each other's learning.
Intrapersonal Skills self-discipline, **independent** learning, flexibility, adaptability, self-awareness, **perseverance,** self-motivation, compassion, integrity, risk-taking, self-respect, sense of belonging	The project is challenging and requires students to demonstrate perseverance when searching for a project or idea to investigate, and then when they researched the issues around the topic with empathy.
Global Citizenship awareness, tolerance, openness, respect for diversity, intercultural understanding conflict resolution civic / political participation, **respect for the environment,** national identity	The app has a platform that enables the projects to be shared. Many of the projects focused on the design of solutions or strategies to address environmental issues.
Media and Information Literacy accessing information, **locating information, communicating ideas,** analysing information evaluating information and media content	Using the app to collect and then to share the project.

Finally PBL and STEMinist Discover App

An online app, *STEMinist Discover,* was developed around the design thinking process and the United Nations SDGs (https://qnawebapp20211112030527.azurewebsites.net, Image 9). The online app scaffolds pre-service teachers to learn how to implement and facilitate design thinking and inquiry learning; and is available as a platform for wide range of classroom students to collect information, reflect on their knowledge, and communicate with others. The use of an online platform allows students to collaborate from anywhere - locally, nationally, and internationally; and it has been trialled and tested several times during the beta testing. Through this educational innovation, children become the drivers and champions of their learning, co-constructing knowledge alongside their teachers, community members and industry to make real-world impacts on local issues. The online app supports teachers to be facilitators of learning, whereby they support students to identify relevant issues, consider multiple solutions, before taking some form of real action that has impacts on others around them. The app is designed for both face-to-face and online learning and helps facilitate the professional development and implementation of ground-breaking education that pushes towards the 4th revolution of education disrupting the 'four walls of a classroom approach', to allow for learning that is more connected and powerful.

CONCLUSION

The use of innovative technology, seen both in the Makerspace and the Problem-based learning units, designed for both face-to-face and online learning, facilitates teachers' professional development and implementation of ground-breaking education that pushes forward toward the 4th revolution of education.

To facilitate the purposeful creation of STEM capabilities that our world will rely so heavily upon requires teachers' innovation, creativity, and implementation of 'constructivist' and 'constructivist' pedagogical approaches. Through the facilitation of these educational innovation opportunities, children become the drivers and champions of their own learning, co-constructing knowledge alongside their teachers, community members and industry to make real-world impacts on local issues. The nurturing of children who believe they can make a real difference to this world is one of the most significant impacts that education can have. It develops their voice, agency, and empathy to see issues from multiple perspectives, which creates critical and valuable self-awareness (Vuorikari, et al., 2016). The development of international learning communities has the potential to influence and change the education system. Facilitating learning through the lens of the United Nations 2030 Sustainable Development Goals positions students to develop a holistic understanding of real-world issues, where they are given opportunities to develop true empathy, critical-thinking and active voice. By building upon this premise and connecting these communities of learning locally and globally students learn about the current world issues and their possible solutions.

The nurturing of children who believe they can make a real difference in this world is one of the most significant impacts that education can have. It develops voice, agency, and empathy to see issues from multiple perspectives, which creates critical and valuable self-awareness. This has been incredibly powerful for both the students and the communities which have been impacted by their involvement. Industry 4.0 should instil knowledge and skills that would not usually be imparted in schools, leading to greater confidence, potential employability and transformation in students by turning them into lifelong learners.

REFERENCES

Boston Consulting Group. (2022, May 23). *37% of the Top 20 Skills Requested for the Average US Job Have Changed Since 2016*. BCG Global. https://www.bcg.com/press/23may2022-top-20-skills-requested-for-average-us-job

Broadley, K. (2015). Entrenched gendered pathways in science, technology, engineering and mathematics: Engaging girls through collaborative career development. *Australian Journal of Career Development, 24*(1), 27–38. doi:10.1177/1038416214559548

Cunningham, C. M., & Carlsen, W. S. (2014). Unpacking its definitions: A conceptual analysis of integrated STEM education. *Journal of STEM Education: Innovations and Research, 15*(1), 37–48.

Deloitte. (2017). *Soft Skills for Business Success*. Deloitte. https://www2.deloitte.com/au/en/pages/economics/articles/soft-skills-business-success.html

Dewey, J. (1902). The Elementary School Teacher. *The school as social centre, 3*(2) (73-86). https://www.journals.uchicago.edu/doi/pdf/10.1086/453152

Knopke, V. (2019). Where are all the students? Factors that encourage female participants in technology education. In P. Williams & D. Barlex (Eds.), *Explorations in Technology Education Research. Contemporary Issues in Technology Education* (pp. 223–239). Springer. doi:10.1007/978-981-13-3010-0_15

Koul, R., Sheffield, R., & McIlvenny, L. (2021). *Teaching 21st Century Skills*. Springer. https://link.springer.com/book/10.1007/978-981-16-4361-3

Koul, R. B., Fraser, B. J., Maynard, N., & Tade, M. (2017). Evaluation of engineering and technology activities in primary schools in terms of learning environment, attitudes and understanding. *Learning Environments Research*. doi:10.100710984-017-9255-8

Larmer, J., Mergendoller, J., & Boss, S. (2015). *Setting the standard for project-based learning*. ASCD.

Lewis, T. (1998, Fall). Petrina, s. and Hill, A. M. (1998) Problem Posing-Adding a Creative Increment to Technological Problem solving. *Journal of Industrial Teacher Education, 36*(1). https://scholar.lib.vt.edu/ejournals/JITE/v36n1/lewis.html

Makrides, G. A. (2019). The evolution of education from Education 1.0 to Education 4.0: Is it an evolution or a revolution? *Cloud*. https://www.l-cloud.eu/wp-content/uploads/2019/03/Evolution_of_Education.pdf

McKinsey Global Institute. (2017). *Skill Shift: Automation and the Future of the Workforce*. McKinsey. https://www.mckinsey.com/featured-insights/future-of-work/skill-shift-automation-and-the-future-of-the-workforce

Organisation for Economic Co-operation and Development (OECD). (2020). *Back to the Future: Four scenarios for schooling*. OECD. https://www.oecd.org/education/back-to-the-future-s-of-education-178ef527-en.htm

Papert, S. (1980). *Mindstorms. Children, Computers and Powerful Ideas*. Basic books.

Pethrick, H., Nowell, L., & Oddone Paolucci, E. (2017). Psychosocial and career outcomes of peer mentorship in medical resident education: a systematic review protocol. *Systematic Reviews, 6*. https://www.ncbi.nlm.nih.gov/pubmed/28859683

Piaget, J. (1936). *Origins of intelligence in the child*. Routledge & Kegan Paul.

Rand Corporation. (2012). Teaching and Learning 21st Century Skills: Lessons from the learning sciences. Asia Society. http://asiasociety.org/files/rand-1012report.pdf

Robinson, V. M. J. (2010). From Instructional Leadership-to-Leadership Capabilities: Empirical Findings and Methodological Challenges. *Leadership and Policy in Schools*, *9*(1), 1–26. doi:10.1080/15700760903026748

Sanders, M. E. (2015). *The Original "Integrative STEM Education" Definition: Explained*. Vtech Works. https://vtechworks.lib.vt.edu/handle/10919/51624

Smith, A., Hielsher, S., Dickel, S., Soderberg, J., & Oost, E. (2013). Grassroots Digital Fabrication and Makerspaces: Reconfiguring, Relocating and Recalibrating Innovation? (University of Sussex, SPRU Working Paper). https://papers.ssrn.com/sol3/papers.cfm?abstract_id=2731835

Stanford Design School. (2014). *Hasso Plattner Model Stanford design school*. Stanford Design school; Stanford Design School. https://dschool.stanford.edu/resources

Tan, S. Y., Al-Jumeily, D., Mustafina, J., Hussain, A., Broderick, A., & Forsyth, H. (2018). *Rethinking Our Education to Face the New Industry Era. EDULEARN18 Conference*, Palma, Mallorca, Spain. 10.21125/edulearn.2018.1564

Timms, M., Moyle, K., Weldon, P., & Mitchell, P. (2018). Challenges in STEM learning in Australian schools. *Policy Insights Issue 7*. Camberwell, VIC: ACER. https://research.acer.edu.au/policyinsights/7/

Tschimmel, K. (2012). *Design Thinking as an effective Toolkit for Innovation*. In: *Proceedings of the XXIII ISPIM Conference: Action for Innovation: Innovating from Experience*. Barcelona. ISBN 978-952-265-243-0.

United Nations Educational, Scientific and Cultural Organization (UNESCO). (2015). *Asia-Pacific Education Research Institutes Network (ERI-Net) regional study on transversal competencies in education policy and practice (Phase I): Regional synthesis report.* UNESCO. https://unesdoc.unesco.org/ark:/48223/pf0000231907

Vuorikari, R., Ferrari, A., & Punie, Y. (2019). Makerspaces for Education and Training: Exploring future implications for Europe. *Ideas.* https://ideas.repec.org/p/ipt/iptwpa/jrc117481.html

Vuorikari, R., & Punie, Y. Yves, Carreton, S., Van den Brande, L. (2016). *DigComp 2.0: The Digital Competence Framework for Citizens.* European Commission. https://publications.jrc.ec.europa.eu/repository/bitstream/JRC101254/jrc101254_digcomp%202.0%20the%20digital%20competence%20framework%20for%20citizens.%20update%20phase%201.pdf

World Economic Forum. (2016). *What is the fourth industrial revolution?* WEF. https://www.weforum.org/agenda/2016/01/what-is-the-fourth-industrial-revolution/

World Economic Forum. (2020). *Mapping TradeTech: Trade in the Fourth Industrial Revolution.* WEF. https://www3.weforum.org/docs/WEF_Mapping_TradeTech_2020.pdf

Wrigley, C., & Straker, K. (2017). Design Thinking pedagogy: The Educational Design Ladder. *Innovations in Education and Teaching International*, *54*(4), 374–385. doi:10.1080/14703297.2015.1108214

APPENDIX

Figure 4. The bee hotel

Figure 5. All creative ideas on what to make with a 1.25 L bottle

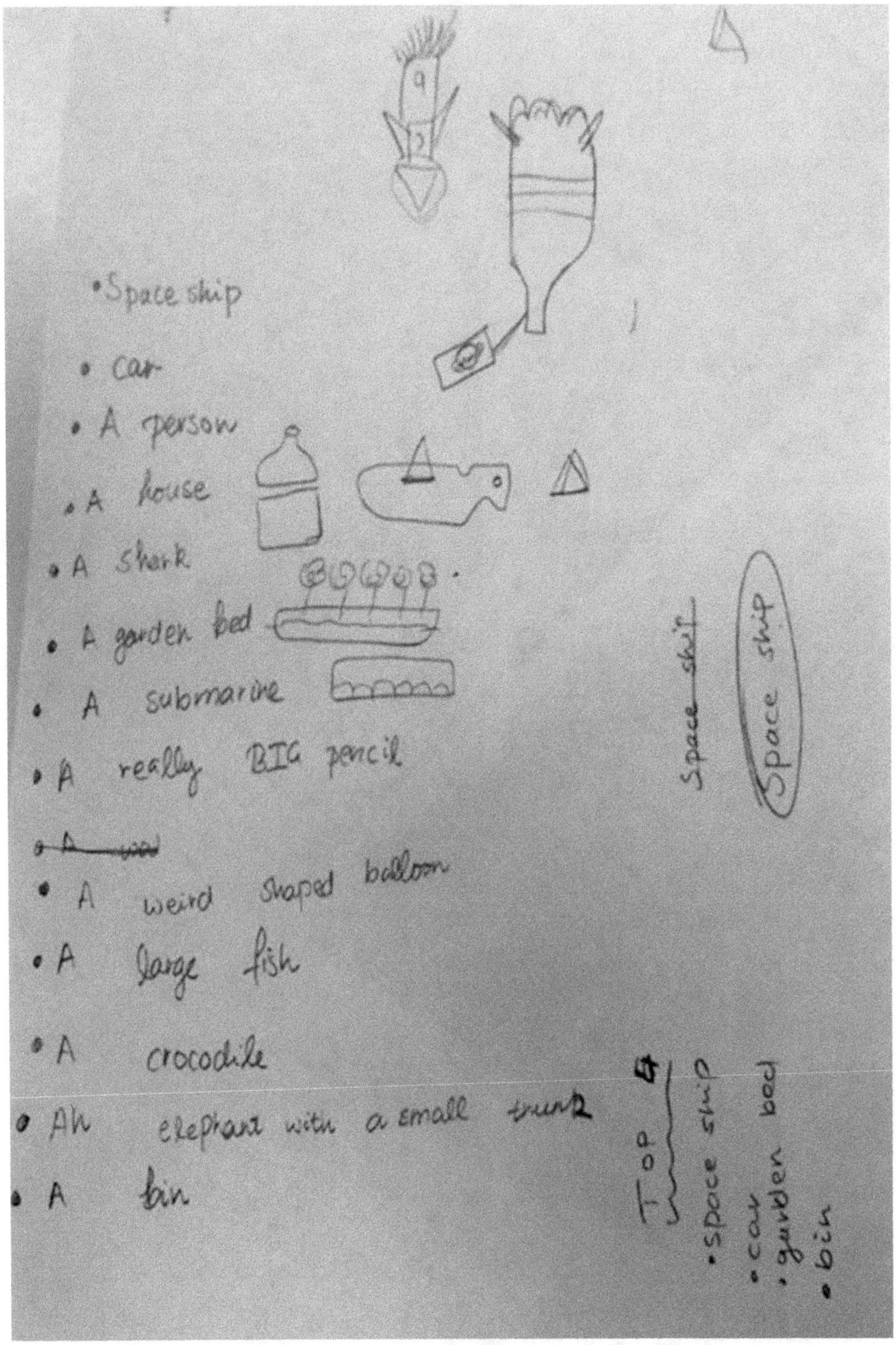

Figure 6. Stomp rocket

Figure 7. Wigglebots

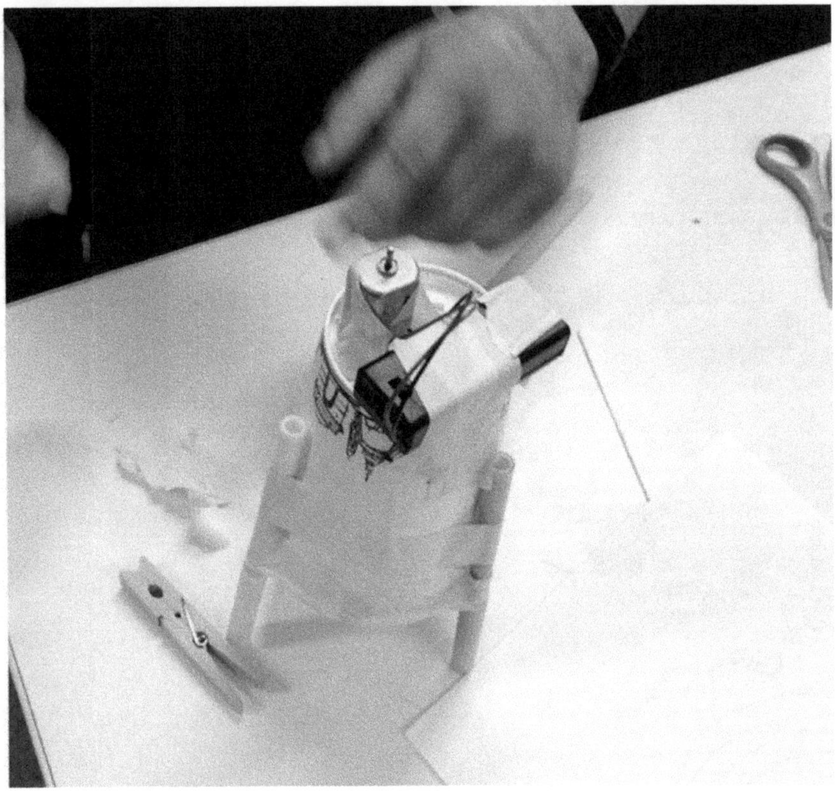

Figure 8. Grabber

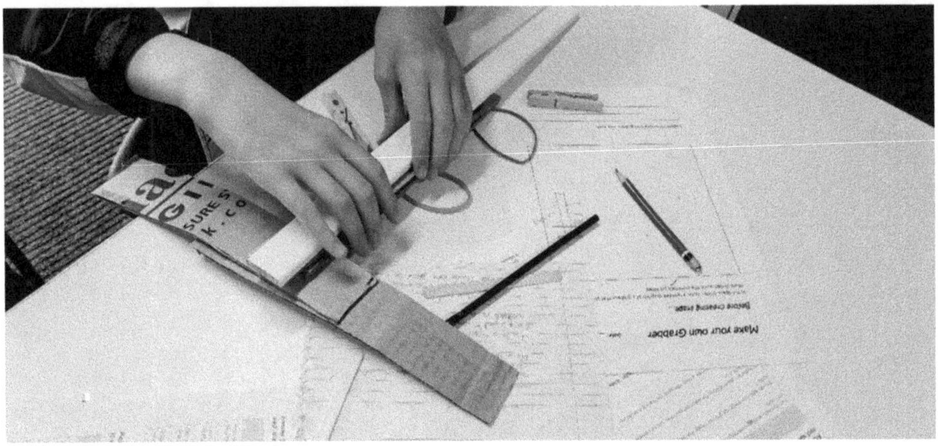

Figure 9. Example of website 1 showing the steps of the model

Figure 10. Example of website with student example (discovery of the 6D model)

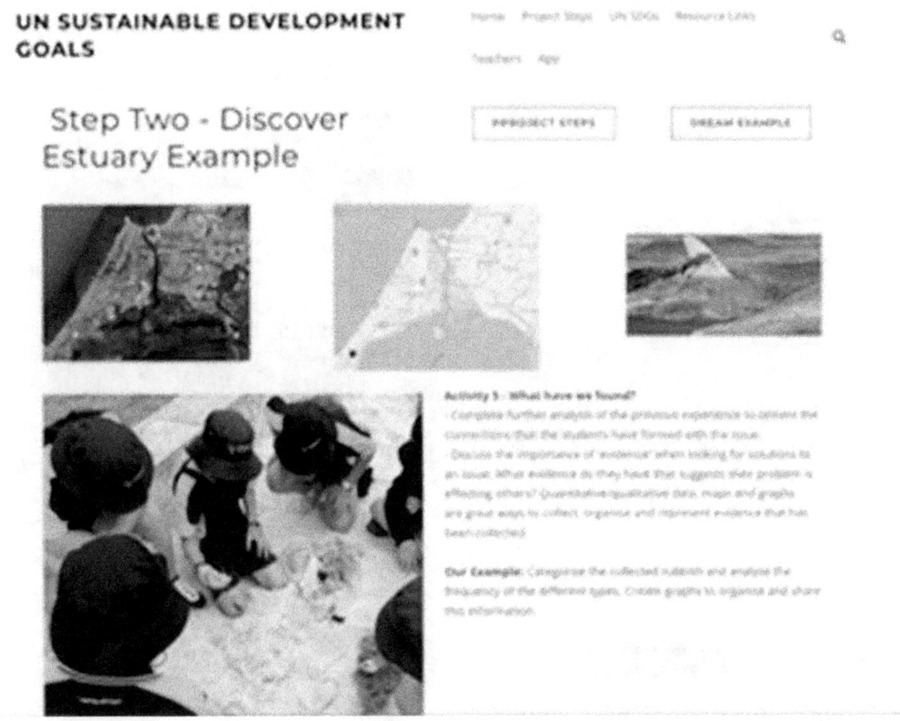

Figure 11. Example of the app using the 6D model

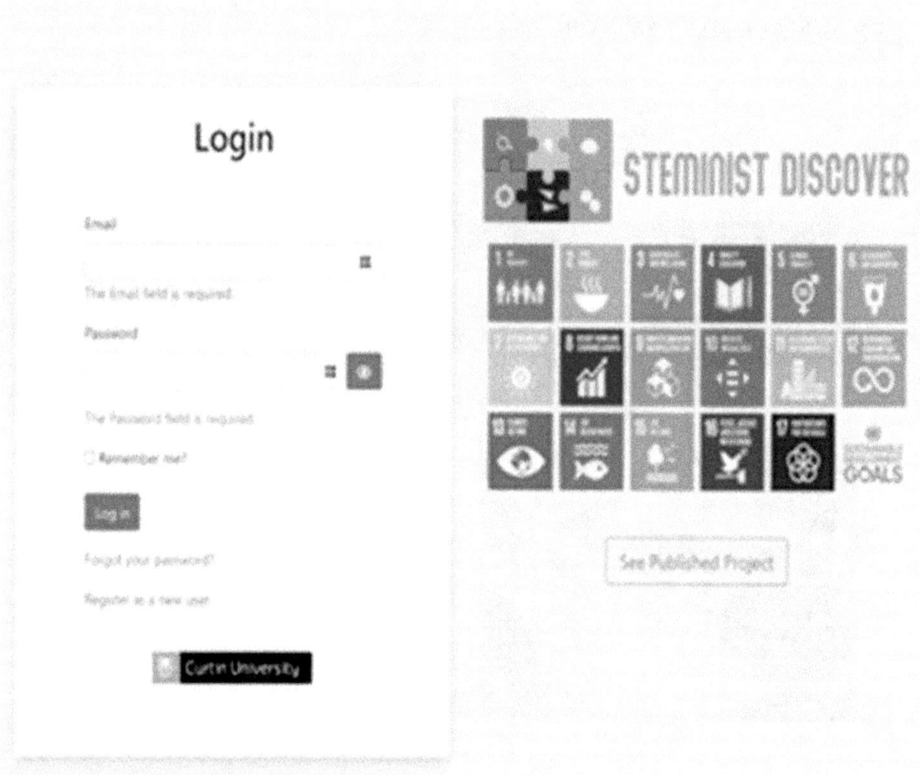

Chapter 9 IBSE Inquiry-Based Science Education: An Example of Application in a Primary School

Federica Lizzi
Università degli studi dell'Aquila, Italy

ABSTRACT

In this study the authors shall endeavour to find answers to how the development of scientific competencies can be taught to primary school children and which methods can be adopted in order to make science learning meaningful according to an inquiry-based approach in a primary school setting. In section one, the methodology aims at active and meaningful learning, and at developing critical thinking and collaborative skills. The teaching methodology based on inquiry will be highlighted showing how this approach is stimulating and motivating for students, enabling them to apply their research skills, to build on meaning and to acquire scientific knowledge through practical experience. In section two, a programme followed by a third grade class will be described. The learning of science actively involves the students in its different stages - formulating the research questions, planning, and carrying out the research to answer the question all are stimulating activities. The following is a simple example of the 5E methodology, an approach based on inquiry.

DOI: 10.4018/978-1-6684-6932-3.ch009

"If I thought of a future, I dreamt of one day founding a school in which young people could learn without boredom, and would be stimulated to pose problems and discuss them; a school in which no unwanted answers to unasked questions would have to be listened to," (Karl Popper, Unended Quest, p.40.)

INTRODUCTION

Competence in science refers to the ability and willingness to explain the natural world by making use of the body of knowledge and methodology employed, including observation and experimentation, in order to identify questions and to draw evidence-based conclusions. Competences in technology and engineering are applications of that knowledge and methodology in response to perceived human wants or needs. Competence in science, technology and engineering involves an understanding of the changes caused by human activity and responsibility as an individual citizen. (The Council of the European Union, 2018, p.9)

To be able to analyze, apply scientific thought, understand the interaction between nature and the world created by man, to maintain a critical approach regarding the validity of information are all competences necessary to each one of us in this world today. (European Commission, 2022, p.3)

Scientific disciplines, therefore, facilitate the development of critical and creative thinking providing those competences necessary to become responsible and active citizens in a society in which science and technology play a role. Thus, scientific literacy is regarded as a precise key in PISA[1] 2015, as "as the ability of an individual to approach types of scientific questions with ideas that concern science as a citizen capable of reflection". For this reason, the teaching of science is compulsory in schools in many European countries as a supplementary subject for at least part of primary education.[2]

However, in spite of the emphasis given to scientific education, international research demonstrates a rather critical reality:

- "On average across OECD countries, 78% of students attained Level 2 or higher in science. At a minimum, these students can recognize the correct explanation for familiar scientific phenomena and can use such knowledge to identify, in simple cases, whether a conclusion is valid based on the data provided.

- On average across OECD countries, 6.8% of students were top performers in science in 2018, meaning that they were proficient at Level 5 or 6" (OECD, 2019, p.112)

Italian students, in particular, have achieved an average science result below the average OECD in PISA testing (Italy 468 vs OECD 489) while the percentage of fifteen year olds does not reach the basic level of 29,9% (18,2% level 1A; 6,6% level 1B; 1,1% below level 1B).

At Level 1A, students can use common content and procedural knowledge to recognize or identify explanations of simple scientific phenomena. With support, they can undertake structured scientific enquiries with no more than two variables. They can identify simple causal or correlational relationships and interpret graphical and visual data that require a low level of cognitive ability.

At Level 1B, students can use common content knowledge to recognize aspects of simple scientific phenomena. They can identify simple patterns in data, recognize basic scientific terms and follow explicit instructions to carry out a scientific procedure. (p.117)

These results should lead to a close reflection of the quality of scientific education and the need to renew teaching methods beginning right from the Primary School. Teachers at this scholastic level are in fact responsible: at the initial stage, the teaching methodology based on IBSE was analyzed. The approach revealed stimulating and motivating results for the students in that they were able to apply research skills, acquire meaning and scientific knowledge through tangible experience. At the Primary School Level Grade Three, an example of the IBSE method was introduced in the second stage.

IBSE: Inquiry Based Science Education

IBSE Teaching

In response to the need to reform teaching skills in science and improve the learning acquisition of students, the Inquiry-Based Science Education (IBSE) was developed. In the Rocard Report of 2007, the European Commission recommended a similar methodology at both the primary and secondary levels to involve students in an active learning process. "A reversal of school science-teaching pedagogy from mainly deductive to inquiry-based methods provides the means to increase interest in science". (p. 2)

Most researchers in teaching consider Inquiry-Based Science Education (IBSE) to be the most effective teaching strategy, because it is a flexible and dynamic methodology, which promotes a student-centered education. The students, who are personally involved, are addressed/oriented/directed to the investigative activity, and guided by the teacher in the conceptualization phase in the formulation of questions and hypotheses followed by the phases of investigation, exploration, experimentation and interpretation of the data resulting from the investigation. In fact, students follow methods and practices similar to those of professional scientists in order to build on their own knowledge. In 1910, Dewey had already affirmed that enacting inquiry activities in science classrooms is beneficial for student learning as it places a strong emphasis on the learners' development of scientific thinking, thus supporting an active approach that, by stimulating students' curiosity they are encouraged to reach a comprehension of concepts on their own and therefore assume responsibility for their learning. Learning based on inquiry, affirmed Lee et al. (2004), promotes student learning through their own research or with guidance; in other words, the teacher asks questions and provides the clues to solve the problem at lower levels, eventually leading to independent inquiry in which it is the students who ask the questions and work out how to solve the problem.

Through the direct experience of natural phenomena, this study represents the heart of learning according to the IBSE model because:

- direct experience is the key that facilitates access to comprehension.
- through direct experience, the students build up their understanding of the world while activities in the class must necessarily begin from their representation of reality.

Conceptions of reality that students elaborate outside the scholastic context can be more or less valid. If this is true, through direct experiences in the classroom students will have the opportunity to reinforce these concepts and relate them to other concepts. On the other hand, given that simple words are insufficient to modify pre-existing ideas, it is necessary for students to adapt to new conclusions on their own in a practical fashion as they generally do outside the scholastic environment.

However, direct experience alone is not enough to build up their understanding: students need to reflect, reason, and discuss among themselves according to the activity they are involved in; they must draw up hypotheses, make guesses, experiment, and examine the subject in order to come to conclusions. Practical activity must be included with information gathered from other sources such as books and internet including access to experts.

In contrast to traditional teaching, IBSE teaching stimulates active learning by allowing students to make their own inquiries:

- Students observe phenomena and reality,
- they formulate research questions, significant from a scientific point of view,

- they formulate hypotheses and predict,
- they control the variables,
- they collect experimental evidence,
- they formulate explanations starting from the evidence gathered,
- they evaluate explanations by comparing between peers and with scientific paradigms,
- they argue and discuss the conclusions they have reached.

Teachers in this learning process, no longer have the function of transmitting knowledge because they are the holders of knowledge, but they become facilitators, supporters, and supervisors of their students. The teacher, therefore, creates an environment conducive to investigation, to research by arranging experiences that allow the student to develop ideas, to connect them to his foreknowledge, to share with his peers, to overcome his misconceptions, in short, to develop an awareness and become the creator of his own learning. To summarize, the teacher:

- supports students in planning and conducting the survey,
- provides materials / tools necessary for the research activity,
- guides students in collecting and analysing data,
- supports students in comparing the results obtained with the scientific paradigm,
- promotes discussion and comparison between students by regulating the debate and increasing the acquisition of a more specific language,
- organises individual and group activities until students gradually become more autonomous in the different phases and acquire confidence and an effective working method.

Learning based on inquiry, therefore, as affirmed by Prince and Felder is an inductive approach to teaching and learning, presenting the following characteristics:

1. a student- or learner-centred approach (Kember, 1997) in which the focus of the teaching is on student learning rather than on communicating defined bodies of content or knowledge;
2. active learning is about learning by doing (Gibbs, 1988, Healey & Roberts, 2004) and may involve, for example, students discussing questions and solving problems (Prince & Felder 2006).
3. the development of self-directed learning skills in which students take more responsibility for their own learning;
4. a constructivist theoretical basis (e.g., see Bruner, 1990) which proposes that students construct their own meaning of reality; it is the students who create

knowledge rather than knowledge being imposed or transmitted by direct instruction. (Spronken-Smith, 2008, p.2)

Type of Student Inquiry

The IBSE approach can be applied gradually along a path in which teachers support students in their progression from the lowest level (Structured Inquiry) to the highest one (Free Inquiry), passing through intermediate steps (Controlled Inquiry and Guided Inquiry), as more of their investigative skills mature and they become increasingly more autonomous. This progression is illustrated through the metaphor of learning to swim in a swimming pool (Figure 1).

Figure 1. Type of student inquiry
Note taken from https://www.trevormackenzie.com/sketchnotes

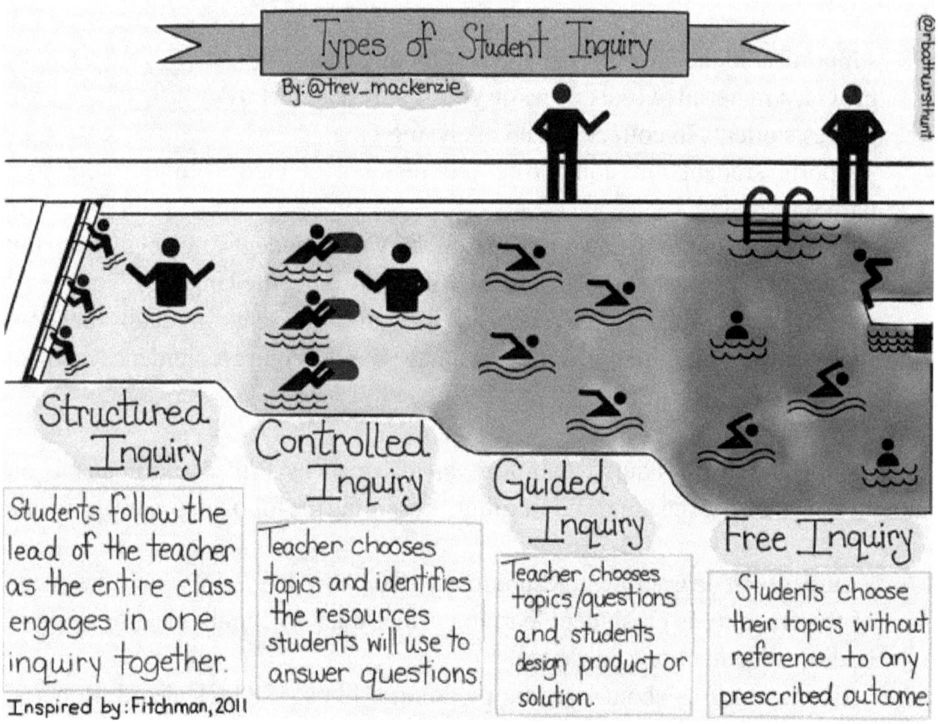

Structured Inquiry

In the shallow end of the swimming pool the swimming instructor is close to the child who cannot swim, while the teacher supports the students, controlling and managing the entire learning process: asking questions, identifying the necessary resources that the students will use to understand, also guiding them in the research phase. The students are like children when they enter the pool for the first time and remain attached to the edge of the pool because they feel insecure – together all of them follow the instructions provided by the teacher.

Controlled Inquiry

Later after becoming familiar with the water, the child lets go of the pool edge but always under the guidance of the instructor who still provides him with the support of the kicking-board. This is the moment when the students begin their first autonomous explorations. Starting from the investigative questions posed by the teacher, they follow the procedures using the resources provided, carrying out the investigations, drawing their own conclusions.

Guided Inquiry

Subsequently the instructor guides the child from the poolside asking him to perform a style. The teacher asks the investigative questions but leaves the students free to select the resources they will use to search for their answers, choosing the procedures to follow.

Free Inquiry

Finally, the now more mature child begins swimming in the deepest part of the pool without the support of the pool-edge, but always under the supervision of the instructor. The students, with the support of the teacher, formulate their investigative questions, research resources and personalize learning activities.

The four types of Inquiry, therefore, differ according to the degree of the students' autonomy in formulating the questions, in planning their research and in drawing conclusions as outlined in the following table (Table 1).

Table 1. Differences between the 4 types of inquiry

Type of inquiry	Who poses the problem	Who indicates the procedure	Who finds the solution
Structured Inquiry	Professor	professor	Professor
Controlled Inquiry	Professor	professor	Student
Guided Inquiry	Professor	Student	Student
Free Inquiry	Student	Student	Student

In a classroom setting, the Inquiry is carried out in a more or less structured way according to the students' investigative skills but also according to their different learning objectives. The application of the various levels of the IBS model, determines.

- an improvement in attitude towards the sciences
- an increase in interest and curiosity
- an improvement in the understanding of concepts
- the development of a higher level of thinking
- collaboration between students

Resistance and Obstruction to the Introduction of IBSE

Several studies carried out by Minner, Levy and Century (2010) illustrate how teaching strategies that actively involve students in the learning process are more likely to increase conceptual comprehension. Inquiry-based teaching develops a greater understanding of content and scientific concepts. It enhances the development of a positive concept of the nature of knowledge and scientific investigation, stimulating the students' motivation and active involvement in the subject.

Nevertheless, this approach is struggling to convince the teaching profession. There are several reasons why teachers remain attached to a traditional methodology:

Too much time needed/ Lack of resources/Greater need of energy /Topics take too long to come to a conclusion/ Inquiry-based lesson difficult to organise/Risk of not fulfilling the national curriculum (and/or not preparing students for the final examinations/Difficulty to evaluate students' work/Students treated as if I they lack skills-competence/Insufficient textbooks/Teaching staff lack motivation.

In Italy, in particular, based on a research project carried out by TEMI[3], 37% of teachers interviewed showed a concern regarding Time, 20% felt anxious about Class Control, 14% about curriculum limitations, 6% felt they had too little time for personal Preparation, 6% about Hostility and/or difficulty with colleagues. Finally, 17% regarded ISBSE as being inefficient.

Let us focus on the first motivation. It is doubtful that IBSE requires time for didactic preparation activity or for classroom organisation. Similarly, regarding teachers' anxiety concerning the lack of time to complete the set curriculum programme is unsustainable. In reality, the official documents in Italy for the first cycle (National curricula indications in nursery school and the first education cycle) 'offers' a solution to the problem. "In the area of a complex vertical project, teachers may create a programme of activities that allows them to develop basic topics in every scientific sector". (Muir, 2012, p. 66). Teachers may choose topics yearly "plan time and method of instruction which suits them without hurry or superficiality and at the same time stimulates the children to discuss their ideas". (p.66). In the documents for the second cycle (National Indications for high schools, guidelines for technicians and professionals, 2010), what is binding are above all the competences referring to the four Cultural Areas (Area of Languages, Mathematics Area, Scientific-Technological Area, Historical-Social Area) and the key competences of citizenship, all of which cultural areas cover. It is, therefore, necessary "to pay particular attention not only to the extent of the total cultural area, but also to the depth of the content. The alternative is not to teach a lot or a little, or to teach superficial and insignificant notions according to a logical encyclopaedic programme, or according to how much time one has to dedicate to the task; whether it is worthwhile [...] Not everything and immediately, but at least under conditions that are indispensable for developing abilities (that require time), but also a genuine and real scientific culture, a culture that requires the capacity of a logical-abstract way of thinking. (Testoni, 2019, p.30)

The 5E Instructional Model

The 5E model is one of the most widespread versions of the IBSE methodology and is based on a learning cycle in which 5 phases follow one another, Engage, Explore, Explain, Elaborate and Evaluate, hence its name. Introduced for the first time in 1989 according to a study carried out by the Biological Science Curriculum Study (BSCS); it was intended to provide guidance for the science curricula in elementary schools.

The 5E didactic model elaborated by Bybee and Landes, is based on the learning constructive theory. It sees the student as an active part in the constructive process of knowledge; understanding and thinking over activities, students can relate new knowledge to previous ideas. Besides, this model promotes collaborative learning; students work together to solve problems and investigate new concepts, asking questions, observing, analysing, and drawing conclusions. Bybee (1997) indeed declares that using this approach, students redefine, reorganize, elaborate, and change their initial concepts through self-reflection and interaction with their peers and their environment. Learners interpret objects and phenomena and internalize those interpretations in terms of their current conceptual understanding (p. 176).

According to this model, the teacher plays the role of director, setting up the learning activities as well as facilitator, guiding the students in the learning of new concepts.

The 5E model, introduced above, consists of five phases, Engage, Explore, Explain, Elaborate and Evaluate (usually listed in serial order). However, the 5E model is a very flexible and dynamic cycle. The teacher can move freely back and forth within the cycles, 'exploring' and 'explaining' phases before the students are ready to move on to the 'elaborating' phase. Most importantly, the 'Evaluate' phase does not only represent the final stage but accompanies teachers and students during the whole of the educational course. (Figure 2)

Figure 2. The 5E instructional model
Note taken from https://files.eric.ed.gov/fulltext/EJ1058007.pdf

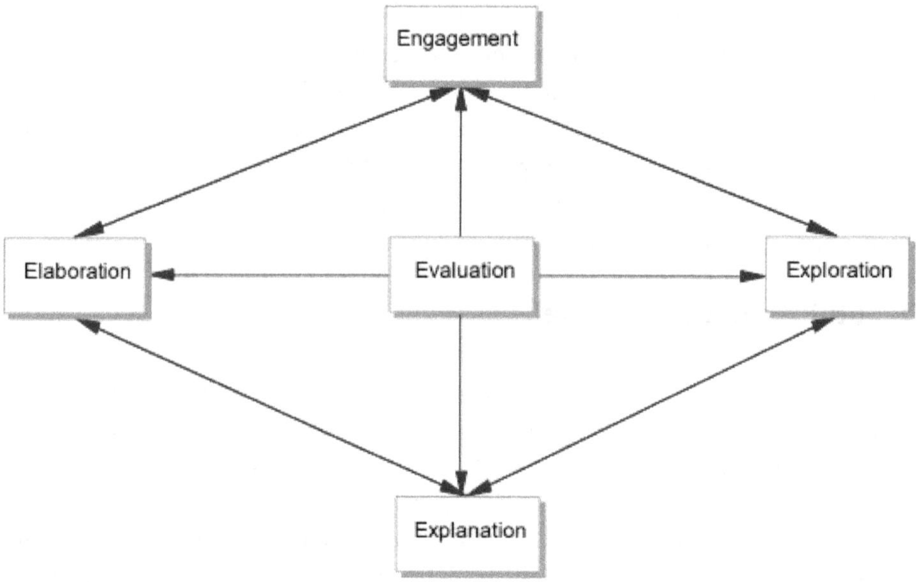

Following this, a table illustrates the five phases of the method, setting out the objectives of both the teachers and the students.

Table 2. Engage

ENGAGE First contact with the topic		
Purpose	**The teacher ...**	**The student...**
– Creating interest and stimulate curiosity. – Setting up learning in a meaningful context – Stimulating the formulation of searching questions. – Creating links with students' previous experiences and knowledge – Bringing out the presence of any misconceptions	– Generates curiosity about the topic – Asks open questions – Listens to students to discover what they already know about the topic – Encourages students to explain their thoughts – Makes connections with reality / students' interests – Avoids explaining concepts, provides definitions and solutions, states conclusions	– Asks questions – Expresses interest about the topic – Shares thoughts / ideas – Answers questions highlighting the knowledge already in his possession.

Table 3. Explore

EXPLORE Concrete learning experiences		
Purpose	**The teacher ...**	**The student...**
– Preparing experiences that allow the students to identify and develop concepts, processes and skills	– Allows students to work together to explore a topic / problem – Asks questions about what they are doing – Listens to the students' ideas while they are talking to each other – Supports the students as needed (without giving answers) – Acts as a facilitator	– Collaborates and shares ideas – Looks for alternative solutions and discusses them with other classmates – Formulates hypotheses and verifies them – Records his own ideas and observations as well as those of the group – Asks questions

Table 4. Explain

EXPLAIN Discussion of the first data that emerged and theoretical connection		
Purpose	**The teacher ...**	**The student...**
– Offering students the opportunity to demonstrate their understanding and knowledge – Offering teachers the opportunity to introduce new concepts, processes or skills	– Encourages students to explain concepts in their own words – Highlights important ideas supplied by students – Asks questions that help students to be precise in their explanations – Helps to organize ideas – Introduces specific words and definitions	– Explains ideas, concepts or possible solutions to other classmates – Uses information derived from explorations to explain concepts – Listens to ideas of others – Asks questions – Thinks over his own learning while elaborating explanations

Table 5. Elaborate

ELABORATE Elaboration of newly acquired knowledge and subsequent application in new situations		
Purpose	**The teacher ...**	**The student...**
– Extending students' understanding and conceptual skills. – Developing a deeper and broader understanding	– Prepares new experiences that deepen the students' understanding. – Asks students to apply content in a new situation or to test additional ideas – Encourages students to extend the concepts of the lesson – Asks students to use the new vocabulary appropriately – Consolidates connections with the real world	– Uses new information to explore additional ideas – Uses new words / concepts to revise explanations – Collaborate to refine ideas

Table 6. Evaluate

EVALUATE Formative feedbacks, formative and summative evaluation, self-evaluation		
Purpose	**The teacher ...**	**The student...**
– Supplying formative feedback to students – Offering students the opportunity to revise and think over their learning and new understanding and skills. – Evaluating student's progress – Encouraging students to evaluate their understanding and skills	– Asks questions that give information about the student's progress – Observes students during the activities – Verifies understanding – Verifies knowledge and skills	– Answers questions that explain or show his own thought – Demonstrates the understanding of the topic while completing activities – Evaluates his own progress and knowledge

An Example of Application in a Primary School

In the final part of this introduction, the focus of this study based on the 5E model is recorded: the course on mould carried out by a grade three class in a primary school.

There are two sections describing each phase of the model:

- in the first section information regarding the project is presented in table form
  - Estimated time
  - Description of activity
  - Teacher's actions
  - Students' actions
  - Possible questions to ask students (actual questions)
  - Necessary resources
- in the second section the tasks carried out during the lesson are recorded and documented in a descriptive form according to didactic-methodology.

Table 7. Module

MODULE: Mould		
SCHOOL: primary school	**TEACHER: Federica Lizzi**	**SUBJECT: Sciences**
CLASS: Grade 3	**Number of STUDENTS: 16**	**NECESSARY TIME: 2 months**
AIM: Understanding the conditions that cause the development of Mold		
TARGETS: • Formulating investigative questions • Formulating hypotheses and previsions • Linking a new experience to a well-known experience • Collaborating with others • Communicating one's own ideas • Providing evidence to support one's own arguments		

Table 8. Stage one: ENGAGE

ENGAGE		
Estimated time: 4 hours		
Description of the activity: The students examine some containers with food covered with a layer of mould, think over and identify investigative questions.		
What the teacher does	**What the student does**	**Possible questions to ask the students (effective questions)**
The teacher divides the students into groups of 4 and assigns a role to each member of the group.	The students divided in groups sit at the desks arranged in islands.	
The teacher gives a container to each group. In each container there is food covered with a layer of mould. The teacher gives no information about the contents and invites the students to make their observations.	The students observe what the teacher has given them.	
The teacher goes around the desks, listens to the conversations among the members of the group and asks open questions to check their foreknowledge.	The students discuss among themselves, answer the questions and report their observations in the group exercise book.	– What is in the container? – What is on the food? – Have you ever seen mould? – In which environment (kitchen, bathroom, cellar...)? – On which surface did it develop? – Why does mould develop? – What is mould?
The teacher invites the students of each group to present their observations to the rest of the class and discuss them collectively. The teacher asks open questions.	Students show their containers, talk about their observations and compare.	– Are they all the same? – What are the similarities and what are the differences?
The teacher invites the students to identify investigative questions they will have to answer in laboratory experiences afterwards.	First the students sit in their small group and then in the big one to discuss and write investigative questions on the board with the support of "The question machine"	– Which research or investigative questions can we formulate to discover further information about moulds? – Do the questions you wrote reflect the characteristics of "The question machine"?
Necessary resources: 5 containers with food covered with a layer of mould "The question machine" (see Appendix B – Figure 4) Exercise-book to report observations Digital board to report the investigative questions		

In this first stage, called ENGAGE, the teacher introduces the subject with an activity. In this case the direct observation of the mould present on some food, with the aim of not only increasing students' participation, attracting their attention and stimulating their curiosity, but also of encouraging them to reflect and ask themselves questions, of creating links with their previous knowledge in order to bring out any erroneous conceptions.

Before starting the activity, the teacher prepares the classroom environment by arranging the desks in islands in order to facilitate exchanges of information within the group. Cooperative participation has characterized the whole course of the activity in the belief that learning is not an individual process, but a process of socializing, requiring interaction and collaboration. During the group activities, students not only simulate scientists' work, sharing ideas, discussing and reflecting about what to do and how to carry out the investigation, but they also gain and strengthen soft skills and social competences.

Setting up the group is a sensitive moment of the didactic planning experience. After considering each student's individual characteristics, the teacher makes four heterogeneous groups selecting four members in each group, assigning each member a specific role and task:

- Reporter: reports the group's reflections to the class;
- Secretary: records the observations in the group's exercise-book;
- Moderator: coordinates the interventions of the group's members;
- Time manager: controls the duration of activities.

After setting up the environment and allocating the different tasks, the teacher gives each group a container with different types of food covered with a layer of mould (Table 9):

Table 9. Type of food

GROUP A	GROUP B	GROUP C	GROUP D
Apple	Lemon	Bread	Grape

The task consists in observing and describing what is in the container and registering the group's considerations. A considerable amount of time is allocated for the student's remarks and / or among students as well as open questions asked by the teacher.

The questions, aimed at initiating a discussion, have the effect of strengthening the focus of learning and of making the students' foreknowledge emerge (see Appendix A-Figure 3). The teacher asks some questions in the planning stage, while the students intervene with other questions.

The teacher circulates among the groups, listens to the conversations and asks questions.

GROUP A

Teacher: What's in the container?
Luca: well, there's a bad apple!
Teacher: Why do you say that it's bad?
Davide: Look teacher, there is mould on it
Sara: Mum throws away things with mould!
Teacher: Why does she throw them away?
Sara: Because they are bad and you can't eat them.
Teacher: So in your opinion, food with mould isn't edible?
Morena: It can't be eaten, it's bad for you!
Teacher: But why is there mould?
Luca: When it's rotten, it goes mouldy.
Sara: When it starts stinking.
Luca: Once I forgot the sandwich in my backpack and after a few days there was mould on it.
Teacher: Where did you see the mould?
Morena: In my bedroom, behind the wardrobe there's mould.
Teacher: Why did it go mouldy just behind the wardrobe?
Morena: I don't know. Dad said he had to wash the wall to take it off because it can hurt you.

All the students in the class know mould, they have seen it several times on food that's gone off; so, they associate it with something harmful that cannot be eaten. Only one little girl had a different experience: mould could even be found on walls.

In this case, mould is also regarded negatively. As for its origin, all the children agreed that mould was a consequence of food going off.

After a moment of working in small groups, the whole class gathers around the teacher who leads the large group as they exchange their findings with each other. Each reporter from each group must explain what is in their given container and report his/her group's considerations. Again, as in the small working group, on this occasion too, the whole class are able to discuss freely.

Riccardo: Teacher, but these moulds aren't all the same.

Beatrice: Yes, look the lemon mould is green instead the grape mould is white.
Riccardo: The apple mould is a bit white, too.
Lorenzo: The bread mould is a bit grey.
Teacher: So, the moulds have different colours. Do you notice anything else?
Luca: Our [apple] is spotty.
Federica: Our grape is hairy.
Lorenzo: Our bread is hairy too!
Beatrice: The mould on the lemon looks like powder.
Teacher: Morena, is the mould on the wall in your bedroom like these moulds?
Morena: No teacher, that mould is dark grey and if you touch it, something like a powder remains on your hands.
Teacher: Why are moulds different?
Morena: Because they are on different foods.
Sara: Yes, if it develops on bread, it's hairy and grey, if it is on a lemon, it becomes green and has very small granules.
Teacher: But, have these moulds anything in common?
Federica: some have the same colour, others don't.

It was only on the base of direct observation that the students identified similarities and differences among moulds; so, they focused on colour and shape.

Still in groups, the teacher asks the students in the last part of the activity to write a list of investigative questions that they will have to answer later during the laboratory experience.

The IBSE method starts with the formulation of scientifically significant questions that the teacher makes up in the guided inquiry, or the students make up in the open inquiry. A good research question gives indications about setting up research. "A good question is the first step towards its answer, it's a problem for which there's a solution, it is a challenging question that invites one to take a careful look to a new experiment or a new activity" (Jos Elstgeest, 1985, p.37).

Furthermore, Rothstein and Santana (2013) consider the formulation of questions made by students as the essential skill for learning. It is, therefore, important that students learn how to formulate good investigative questions.

The experience described is an example of open inquiry by which the students are actively involved in the formulation of research questions with the support of "The question machine" (Peeters), a flow chart that, through essential criteria, enables one to determine whether a question is investigative or not. When a question does not fulfil one of the criteria, the students must change it.

Dekker and Baren-Nawrocka (2017) have identified some criteria for research questions:

- The question fits the theme. The question must be relevant to the theme in order to research.
- The answer should not be searchable. A searchable question is a question you can look up the answer for (for example fact questions). Research must be carried out in order to discover the answer. This is the intention (The answer to searchable questions can of course be looked up during the project).
- The student does not know the answer to the question yet. The student is curious about the answer and can learn something from the question.
- The question must be singular. Only one question can be researched at a time. Multiple research questions should be split up into single research questions and answered separately.
- The question is concise and clearly formulated so you know exactly what is being researched.
- The question begs whether the research can be conducted during the school time and whether the materials are available. (p.20)

The questions formulated by the students are written in the exercise-book and then transferred to the board online:

Table 10. Questions

GROUP A	GROUP B	GROUP C	GROUP D
How does mould develop? Does mould develop immediately?	Does mould die? What is mould made of?	How does mould develop? Why does mould exist?	Where can mould be found? Do moulds have the same colour?

The moment of sharing with the class is essential. It enables an analysis of the written questions to be made; students can think all together over their "investigative" nature using the question machine. Being able to formulate research questions is a gradually acquired skill; it is not simple at all. In fact, through confrontation with the class, the questions do not reflect the criteria completely:

- some questions require a simple "yes" or "no" answer (Do moulds have the same colour? Does mould develop immediately? Does mould die?);
- other questions are too generalized (How does mould develop?)
- to some others it is not possible to give an answer (Why does mould exist?);
- some can be answered simply by looking up information in books (What is mould made of?)

With the teacher's help, the students change some questions making them more specific and easier to check:

- What are the conditions that cause the development of mould on food?
- What do our moulds have in common besides their size and colour?
- What happens if there is some mould on food after a few days?

Table 11. Stage two: EXPLORE

EXPLORE		
Estimated time: 2 hours for formulating hypotheses and planning experiments; 1 hour for arranging experiments; some days for observation; 2 hours for verifying and sharing.		
Description of the activity: The students plan and carry out experiments to verify the validity of the formulated hypotheses concerning the investigative question "What are the conditions that cause the development of mould on foods?"		
What the teacher does	**What the student does**	**Possible questions to ask the students (effective questions)**
The teacher asks each group to formulate hypotheses about the first research question: "What are the conditions that cause the development of mould on foods?" The teacher walks among the desks, listens to the conversations, observes and asks questions.	The students, always in groups, debate among themselves, answer the teacher's and write their hypotheses.	– What are the possible answers to the research question? – And why do you think this is?
The teacher asks each reporter to share the hypothesis with the rest of the class. Asks questions to promote discussion.	The reporters relate to the class about their work. All the students debate and after they write their hypothesis on the virtual board.	– What do you think about the idea just presented? – Do you agree?
The teacher asks each group to work on one only hypothesis and to plan an experiment to verify its validity. The teacher walks among the desks, listens to the conversations among the student of the group and asks questions.	Students choose the hypothesis, debate among themselves, answer the questions, plan their experiment and write in the "Research Plan"	– What do you think of doing to check if the hypothesis is correct? – What do you need? – How will you carry out the experiment? Describe it – How will you collect the data? – What could happen if...?
The teacher brings the material required by the students to make the experiment.	The students set up the experiments.	
	During the following days, the students, observe and write in the "Working plan".	
Necessary resources: Necessary materials for the experiments Exercise book "Research Plan" (see Appendix C) "Working plan" (see Appendix D) Digital board to write hypothesis		

In this second stage, called EXPLORE, the students have the opportunity to be personally involved in building knowledge, concepts and skills through concrete experience. The students explore the research questions, plan, and make investigations to create new ideas.

The aim of exploration activities is to establish experiences that teachers and students can use later to formally introduce and discuss concepts, processes, or skills. During the activity, the students have time in which they can explore objects, events, or situations. As a result of their mental and physical involvement in the activity, they establish relationships, observe patterns, identify variables, and question events. The teacher's role in the exploration phase is that of facilitator or coach. The teacher initiates the activity and allows the students time and opportunity to investigate objects, materials, and situations based on each student's own ideas of the phenomena. If called upon, the teacher may coach or guide students as they begin reconstructing their explanations. Use of tangible materials and concrete experiences is essential (Bybee et al, 2006, p.9).

In the EXPLORE STAGE, the students learn by "doing" and building meanings, making sense of the world, completing and summarizing the new experiences; in this way they re-build what they already know, they rework mental schemes and previous knowledge. Direct experience, therefore, represents the starting point; it comes before explanation and leads to understanding. The vision of learning as a process is evident:

- active (the person's conscious and responsible role);
- constructive (the process of "balancing" between previous mental structures and new knowledge);
- collaborative (the dynamics of social interaction in which learning develops);
- intentional (the role of motivational and desired processes in the development of potential learning);
- conversational (the role of language when structuring the comparison and negotiation of meanings among actors);
- contextualized (the reference to reality tasks when situating the learning process);
- reflective (the recursive circle between knowledge, experience and reflection of the subject on which the intrapsychic negotiation is structured) (Castoldi, 2012, pp.40-41).

At this stage, it is fundamental to formulate hypotheses correctly and to lay out a research plan that provides for the analysis of a variant at a particular time. This is an acquired skill with practice and above all, with the support of the teacher who, with the right questions, stimulates students' reflection.

Initially, the teacher, in this case, works with single groups and then with the class group to encourage the students to think about possible hypotheses and related research plans.

This document illustrates the process that took place - the single groups and the class group to formulate hypothesis, isolating the different variables that come into play, in planning the research activity.

Teacher: When does mould develop on food?
Morena: When something to eat goes bad.
Teacher: How does it go bad? Why does food go bad?
Lorenzo: Once mum left a mozzarella outside the fridge. It went off and she threw it away.
Teacher: What do you mean by go off?
Lorenzo: It stank and was a bit yellowish
Teacher: Was there mould on it?
Lorenzo: Yes, it was hairy.
Teacher: So, if mozzarella remains outside the fridge, what happens?
Lorenzo: It goes off and mould develops.
Teacher: If it remains in the fridge instead?
Riccardo: It doesn't deteriorate and mould doesn't develop.
Teacher: So in our case, do you think that the mould depends on whether we put food inside or outside the fridge?
Beatrice: Yes yes. If it is inside it doesn't develop, if it is outside, it develops.
Teacher: How do we need to verify this hypothesis?
Riccardo: We must take a mozzarella.
Teacher: And then?
Beatrice: We wait.
Teacher: But where do we put this mozzarella?
Riccardo: Outside the fridge.
Teacher: Let's check the hypothesis. Is putting a mozzarella outside the fridge all we need to do?
Morena: No. We must put a mozzarella inside the fridge too.
Lorenzo: But how? There is no fridge here.
Teacher: I will bring a small fridge. Then we can put one mozzarella in the fridge and one outside the fridge. But what do we put these mozzarellas in?
Beatrice: The one inside the fridge, in a container, and the one outside, on a dish.
Teacher: Why this difference?

Morena: Mum puts things inside in the fridge in closed containers and when she takes them out, she puts them on a dish.

Teacher: We want to verify if the mould development depends on staying or not inside the fridge. If the mozzarella outside the fridge is on a dish, on an open dish, what is the difference between the two mozzarellas?

Riccardo: One is inside and the other is outside.

Lorenzo: One is on an open dish, and one is in a closed container.

Teacher: So, there are two different conditions: inside / outside and open / closed. How can we understand if mould is because of being inside or outside the fridge and not being in an open or closed container?

Lorenzo: Well, we put the mozzarellas in a closed container, so if mould develops on the mozzarella that is outside, it depends on the fridge.

Davide: Teacher, but if we take four mozzarellas, we put two of them inside the fridge and two of them outside, two in an open container and two in a closed one.

Teacher: Explain the situation better. Go to the blackboard and do a drawing.

Davide: These two mozzarellas [he draws them on the board] are inside the fridge, this one [he points to it] is in an open container and this one [he points to it] is in a closed container. And these two [he points to them] are outside. This one [he points to it] is in an open container and this one [he points to it] is in a closed container.

Teacher: So, what could happen? If mould develops ... what does it depend on?

Davide: If mould develops on the mozzarellas outside the fridge, then it depends on being outside the fridge. If it develops on mozzarellas on an open container, then it depends on whether it fits in an open container.

Federica: Teacher, and if it develops only on one mozzarella?

Teacher: What do we have to look at?

Davide: Let's look at which mozzarella it is.

Teacher: And so?

Sara: If it's the one inside the fridge.

Morena: But there are two of them in the fridge.

Sara: The one inside the fridge, inside the open container. So, we can say that mould develops if it is in the fridge in an open container.

Luca: In my opinion it develops on the mozzarella that is outside on the open dish.

Teacher: So, Luca, you expect that mould will develop on the mozzarella in the open container, outside the fridge. What do the others expect?

Most students agree with Luke's prediction.

Because of the students' age and because this is their first experiment, they are constantly referring to their life-experience: mum puts the food inside the fridge, so they don't go off and so mould doesn't develop. At this point, the teacher takes into consideration the children's need to refer to their life experience, home and what Mum does.

The debate and the reflection, therefore, starts from their personal experience. The debate reveals the difficulty in isolating the single variables: the students don't realize that in an experiment they only have to change one parameter at a time, keeping the others constant. In fact, at first they suggest putting a mozzarella closed in a container inside the fridge and another one in an open container outside the fridge. It was thanks to the teacher's questions that they understood that the mozzarella outside the fridge was in two different conditions: outside the fridge and in an open container. This made them reflect on the fact that, if mould developed, they would have found difficulty in attributing the "responsibility" to one factor or to the other. From this arose the need to consider one variable at a time and therefore to use four mozzarellas, as in the synthetic scheme (Table 12):

Table 12. Fridge/container scheme

	Inside the fridge	Outside the fridge
Closed container	Mozzarella A	Mozzarella C
Open container	Mozzarella B	Mozzarella D

After a long debate about the hypothesis, the teacher invites all the students to write in the "Research plan" (see Appendix C) in which, starting from the research question, they write the hypothesis, make previsions, list the needed materials, describe the working stages and the division of tasks.

During the following days, the teacher brings the materials necessary to carry out the research and the students prepare the experiments, start to observe, to collect and analyse data and evidence, to think about what they have noticed and to build models or initial explanations and to write on the Research Plan board.

The board acts as a guide for experience documentation. The elaboration in writing, through texts, drawings, diagrams, and graphs helps the students during the learning process; it requires them to summarise their thinking and tell the others about it. The reading of these tasks, furthermore, allows the teacher to evaluate progress and the quality of their students' thinking.

Table 13. Stage three: EXPLAIN

<table>
<tr><th colspan="3">EXPLAIN</th></tr>
<tr><td colspan="3">Estimated time: 4 hours</td></tr>
<tr><td colspan="3">Description of the activity: The students exchange their opinions; they confront each other and revise their conclusions taking into consideration the scientific explanations. They use the microscope to discover the structure of moulds and acquire a specific language.</td></tr>
<tr><td>What the teacher does</td><td>What the student does</td><td>Possible questions to ask students (effective questions)</td></tr>
<tr><td>The teacher invites the students to verify their hypothesis, to draw conclusions and share them.
The teacher asks questions.</td><td>The students debate, think over the data collected and write their conclusions. They share their conclusions</td><td>– Which statements can you make?
– What evidence have you collected?
– Which links are there between the data collected?
– What led you to this conclusion?
– Is the conclusion reliable?
– What did you learn from this research?</td></tr>
<tr><td>The teacher integrates the students' explanations with hers, introducing the specific terms</td><td></td><td></td></tr>
<tr><td>The teacher proposes the vision of a short video</td><td>The students watch the video and think over the information received.</td><td></td></tr>
<tr><td>The teacher proposes an activity under the microscope to observe the structure of moulds. Asks questions, introduces specific terms and gives explanations</td><td>The students use the microscope, observe, describe what they see and draw it in their exercise book.</td><td>– What do you notice under the microscope?
– What do these moulds have in common?</td></tr>
<tr><td colspan="3">Resources needed:
Moulds
Microscope
Exercise-book
Video: https://youtu.be/8Y827BFjKnU</td></tr>
</table>

The EXPLORATION stage is followed by a stage of "awareness", directed by the teacher that leads to the contextualizing of what emerged from the preliminary research and so to build interpretative models of reality.

The EXPLAIN phase consists of two parts. In the first part, the students communicate what they have learned up until that moment. The teacher acts as a facilitator asking and helping the students to share their explanations and ideas. In the second part, the teacher provides resources and information to support students' learning, introduces scientific concepts and formal definitions, promotes a discussion regarding the misconceptions that emerged in the previous stages, leading the students towards the acquisition of specific language. The students use these resources and information to revise their models.

What is important about this method is that the explanation follows the experience, thus allowing the student to include the latter into a larger picture and to think over the connections between the newly acquired knowledge and their previous knowledge.

During the stage EXPLORE, related to the plan realized and here described, the students had different experiences to verify the two hypotheses: the development of mould depends on the temperature (inside or outside the fridge) and the development of mould outside the fridge depends on the contact with air (in an open or closed container).

In this stage, the teacher encourages students to explain, in their own words, the concepts that emerged by asking them questions to help them give accurate explanations, highlights the important ideas provided by the students and introduces specific words.

Teacher: The other day we asked ourselves a question, "What are the conditions that cause the development of mould on food?" To be able to answer, we carried out some experiments. What are your conclusions?

Matteo: It depends on the open container.

Teacher: What evidence do you have to make this statement?

Matteo: Mould developed on the mozzarella in the open container, while on the one in the closed container there isn't any mould.

Teacher: Why, in your opinion, did it only develop on the mozzarella in the open container?

After some minutes during which the students look at each other perplexed by the question, Morena answers: "Mom says not to leave the things to eat in the air otherwise they go off.

Riccardo: It's true, the air goes in.

Teacher: Then why is it, that mould develops if air goes in?

Luca: Maybe in the air there is mould that floats.

Teacher: But have you ever seen mould floating in the air?

The students answer "no" in chorus.

Luca: There could be some very, very small pieces of mould, such as flower-dust. We can't see it, but it is in the air. It is the flower-dust of moulds.

Teacher: This "flower-dust of moulds" really has another name: spores. Spores are not visible to the naked eye and float in the air.

Morena: So mom is right to tell me to close the containers, otherwise spores enter and mould develops.

Teacher: So we can conclude that the development of mould depends on contact with air. But we had two mozzarellas placed inside open containers. What happened to the other one?

Davide: We saw that there was a spot of mould.

Beatrice: We thought that there would be some more in a few days.

Teacher: But why is there only a "little spot of mould" on the mozzarella on the open container inside the fridge while on the one outside there are so many?

Federica: Because the fridge is cool. That's why we put food in the fridge so it doesn't go bad.

Teacher: But in the end a bit of mould also developed on the mozzarella inside the fridge.

Federica: There is also air inside the fridge and so there are spores.

Beatrice: But the spores take longer to develop mould.

Davide: Spores feel cold and work slowly.

Beatrice: They work harder in the heat.

Teacher: Very well, so we can say that a second factor for developing mould is temperature.

The teacher suggests watching a short film (Scienze-Le Muffe #1: https://youtu.be/8Y827BFjKnU) to highlight further factors causing the mould development: darkness and humidity.

The students write this information in their exercise-books.

During the next meeting, the teacher goes back to the "air" factor again to discuss the knowledge of the structure of moulds together with students.

Teacher: Last time, we understood that the development of mould also depends on the presence of spores in the air. Spores cannot be seen to the naked eye, so what can we use?

Davide: Let's take magnifying glasses.

The students take the magnifying glasses and they realize that it is impossible to see spores in the air and on moulds.

Luca: Teacher, let's use a microscope.

Teacher: But what do we put on the slide?

Luca: Let's put some mould, so we'll see the spores.

Students take turns observing under the microscope the different moulds brought by the teacher.

Sara: We can see some filaments and little balls.

Davide: The little balls look like worm eggs.

Matteo: The filaments look like branches.

Starting with the students' observations and comments, the teacher introduces further information and new words: hyphae, sporangium, spores and mycelium.

As we can see from the above, the explanation begins from what emerged previously to make sense of experiences and to organize knowledge.

Table 14. Stage four: ELABORATE

<table>
<tr><th colspan="3">ELABORATE</th></tr>
<tr><td colspan="3">Estimated time: 3 hours</td></tr>
<tr><td colspan="3">Description of the activity:
The students read some articles and search for information to understand why the expiry date is reported on some products, what happens after the expiry date, and what civil liabilities are there in the event of food fraud.</td></tr>
<tr><td>What the teacher does</td><td>What the student does</td><td>Possible questions to ask students (effective questions)</td></tr>
<tr><td>Starting from a student's observation, the teacher invites the students to take to school some empty packs of products.</td><td>The students take the empty packs of products</td><td></td></tr>
<tr><td>The teacher asks the students, divided in groups, to observe the packs and in particular look at the dates written on their back.
The teacher walks among the desks and asks questions for reflection</td><td>The students observe, debate and write in the exercise- book.</td><td>– How many dates are written on the packs?
– What do they refer to?
– What is the wording next to the date?
– What does the wording mean?</td></tr>
<tr><td>The teacher invites the students to ask themselves questions about the subject "expiry dates" written on the products</td><td>The students, divided into groups, debate, think over and write down the questions on the exercise-book.</td><td>– What would you like to know about the subject "expiry dates"?
– Formulate questions that we will try to answer afterwards</td></tr>
<tr><td>The teacher asks each reporter to share the questions with the rest of the class.</td><td>The reporters present their work to the class. All the students debate and then write the questions on the virtual bulletin board.</td><td></td></tr>
<tr><td>The teacher gives a different article to each group. The teacher invites the students to read it and to look for the answers to the questions. The teacher walks among the desks, observes how they got organized and listens to the conversations among the members of the group.</td><td>The students get organized to do the work. They read the assigned article, look for information, debate and write the answers to the questions.</td><td></td></tr>
<tr><td>The teacher asks each reporter to share the answers with the rest of the class and promotes reflection.</td><td>The reporters present the answers to the class. All the students argue.</td><td>– What do you think about your schoolmate's answer?
– Do you agree?
– What information did you find to support this answer?
– Did you find the answer to each question?
– What questions remained unanswered?
– Did you find other useful information?</td></tr>
<tr><td>The teacher invites them to write down the additional information in the notebook.</td><td>The secretaries write the new information</td><td></td></tr>
<tr><td colspan="3">Resources needed:
Exercise-book
Digital card to write questions
Articles:
• https://www.altroconsumo.it/alIMENTO/fare-la-spesa/speciali/alimenti-scadenze#:~:text=La%20data%20di%20scadenza%20veni,un%20alimento%20deve%20essere%20consumato.
• https://www.fondazioneveronesi.it/magazine/articoli/lesperto-risponde/i-cibi-scaduti-vanno-buttati
• https://ilfattoalimentare.it/termine-minimo-di-conservazione.html
• https://www.laleggepertutti.it/514318_data-di-scadenza-alimenti-normativa</td></tr>
</table>

In the ELABORATE stage, the students apply the learned concepts to new situations or they solve problems in order to deepen and reinforce comprehension to generalize the concepts and to analyse specific matters that have emerged. In this stage, the students expand on the learned concepts, create links with other correlated concepts integrating sciences with other disciplines, and apply their comprehension to the surrounding world in new ways.

About the didactic project illustrated in this chapter, the teacher decides to make a examine EXPLAIN - a matter which came up in the previous stage. During the confrontation with her schoolmates, the student Sara said: "the mozzarella that we used had expired."

Teacher: How do you know?
Sara: There is an expiry date on the boxes.
Matteo: We can eat things until that date and then we have to throw them away.
Walter: But how do we know when they expire?
Teacher: Would you like to discover it? Next time we will just deal with this topic.

The teacher decides to treat the subject about product expiry and, in particular, she wants to go into depth about some essential concepts:

- Difference between expiry date and minimum conservation term
- Food classification:
    - Foods with expiry date,
    - Foods with minimum conservation term
    - Foods with no expiry date and no minimum conservation term
- "Conscious label"
- Sanctions in case of fraud

The activity begins with an observation and reflection task: the students have to read the dates written on the packs.

Federica: Teacher but on the crackers the writing says: "to be consumed preferably within", while on the milk "to be consumed within."
Riccardo: It's true on one there is the word "preferably" and on the other there isn't.
Teacher: What does the word "preferably" mean?
Carlo: I prefer milk, that is I like milk.
Teacher: Why do you prefer milk, besides of the fact that you like it?
Carlo: Because it's good.
Teacher: Milk is good, so should all the children drink it?
Carlo: No, Davide doesn't like it.

Teacher: Then we could say "Children should preferably drink milk", that is we advised it but it's not an obligation.

Carlo: Yes.

Federica: So I'm not obliged to eat the crackers within this date.

Teacher: It is advised but it isn't an obligation, so you can choose to eat or not a pack of crackers after that date.

Riccardo: But on the milk package there isn't the word "preferably", so I can't choose. I can't drink it after that date.

Carlo: But on the sugar package there is no date.

Riccardo: So, it doesn't expire, we can always eat it.

Teacher: For sugar, as for other foods, the expiry date isn't compulsory.

This activity gets children to read labels on the packages carefully - think about information collected and make a first classification among foods: some with an expiry date, others with an advised expiry date and more without a date. Furthermore, this task arouses the curiosity of the students; they begin to ask questions.

The teacher, using the strategy of "anticipatory questions", asks the students to write a list of questions for which they would like to have answers, then the teacher gives four articles, one for each group, and invites them to read and look for the answers to their questions.

Table 15. Questions

QUESTIONS FORMULATED BY THE STUDENTS
• Who sets the expiry date? • Why on some foods there is the writing "to be consumed preferably within" and on others only "to be consumed within"? • What happens if you don't respect the expiry dates? • Can a shopkeeper sell an expired product? • What happens to food when it reaches its expiry date?

The anticipatory questions can be considered "advanced organizers", that is organizational tools for didactics which help students to classify information more easily and so to orient themselves in the topics covered. They represent didactic mediators as they stand between reality and the subject to facilitate their representation.

Formulating questions, even before reading, facilitates the understanding of the text: they serve to form a general idea of the text before reading it, they recall previous knowledge, create curiosity and motivation, but above all they allow a deeper processing of the subject because answering them guides their attention and improves concentration.

During the reading of the articles, the students look for answers to the anticipated questions and identify additional information given in the text.

Table 16. Stage five: EVALUATE

<table>
<tr><th colspan="3">EVALUATE</th></tr>
<tr><td colspan="3">Estimated time:</td></tr>
<tr><td colspan="3">Description of the activity:
The students make a self-assessment of the understanding of what they learnt and of the skills acquired, and the teacher evaluates students' progress in achieving learning goals.</td></tr>
<tr><td>What the teacher does</td><td>What the student does</td><td>Possible questions to ask students (effective questions)</td></tr>
<tr><td>The teacher observes the students during the activities and records ("Observation check list")</td><td></td><td></td></tr>
<tr><td>The teacher provides formative feedback during the activities</td><td></td><td></td></tr>
<tr><td>On the basis of the feedbacks provided by the students, the teacher calibrates the activity</td><td>The students provide feedback for the teacher</td><td></td></tr>
<tr><td>At the end of the activity, the teacher suggests creating a summary map
The teacher evaluates the students' work using the evaluation rubric ("Rubric-Map")</td><td>The students, individually, create a map using their acquired knowledge</td><td></td></tr>
<tr><td></td><td>The students, individually, think and fill in the board "I self-assessment" and "Cognitive autobiography".</td><td></td></tr>
<tr><td colspan="3">Resources needed:
Cards:
• "Observation check list" (see Appendix E)
• "Rubric-map" (see Appendix F)
• "I self-assessment" (see Appendix G)
• "Cognitive autobiography" (se Appendix H)</td></tr>
</table>

The EVALUATE phase, in the sense of assessment, even if included as the last step of the planning, is developed during all the activity, as

The evaluation comes before, accompanies, and follows the curricular activities. It activates the actions to be taken, regulates the actions initiated, promoting the critical balance on those completed. It assumes a prominent training function of

accompaniment of learning processes and a stimulus for continuous improvement (MIUR, 2012, p.13).

CONCLUSION

The *assessment* perspective *for learning* has a formative basis so teachers, according to pre-established standards, collect information and data to adapt their work to the students' needs. The aim is not to judge the work done by the students, but to highlight what they understood and how they organize their knowledge, to prepare the next stage in the teaching / learning process and to provide them with the feedback necessary to achieve the goals set. The evaluation, however, does not concern only the student, but also the teacher who can thus verify the success of the process devised and adapt it to the real needs of the learners.

Furthermore, the formative assessment, in the sense of *assessment as learning* (Earl, 2003), focuses attention on the students' metacognitive processes. The student, placed at the centre of the whole process, is seen as a "critical connector" between assessment and learning. The student, as a critical thinker, gives a sense to information and links them to those previously acquired to build new learning; he uses evaluative feedbacks to monitor and remodel his own learning process asking himself reflective questions and analysing the various learning strategies. Thereby, assessment facilitates the development of the metacognitive and self-regulation skills to encourage the habit of revising and to bring his own knowledge into question, in the light of lifelong learning. In this way self-assessment becomes the focus of the whole process.

The teacher evaluates in progress, taking account of the formative nature of assessment; he is continuously evaluating. He is constantly observing students, collecting evidence during debates, preparing moments of formative feedback and opportunities for self-assessment, inviting the students to think over their tasks and their interaction with their classmates to develop in them a metacognitive awareness and to provide further additional elements for formative assessment.

The teacher can use several tools: rubrics, check list, portfolio, etc. Students can show their comprehension through the elaboration of final products, as for examples newspaper articles, essays, presentations, models, etc.

In the case described in this chapter, the teacher arranged a range of tools:

- "Observation check list" (see Appendix E) to
  - detect the students' behaviour and attitudes, both in group works and in the individual ones,
  - write down the evidence found during the debates

- “Rubric-map” (see Appendix F) to verify if the students
    - identify the main concepts and arrange them consistently
    - use an appropriate terminology
    - identify relationships among the main concepts.
- “I self-assessment” (see Appendix G) to
    - stimulate the student’s reflection about his own capacities (‘I am able to...’), about his role inside the group (‘I’ in the group) and about the group management (‘we’ inside the group)
    - provide the teacher with information about the student’s level of awareness.
- “Cognitive autobiography” (see Appendix H) to stimulate the student’s reflection
    - about the way he worked during the different phases of the task
    - on the difficulties encountered and above all on the strategies adopted to overcome them
    - on what the student learned and how he learned it.

REFERENCES

Bansal, G. & Ramnarain, U. (2021). Inquiry-Based Science Education in Primary Schools. *Education, 49* (3), 259-262. doi:10.1080/03004279.2020.1854955

Bybee, R. (1997). *Achieving scientific literacy: From purposes to practices.* Heinemann Publications.

Bybee, R., Taylor, J., Gardner, A., Scotter, P., Carlson, J., Westbrook, A., & Landes, N. (2006). *The BSCS 5E Instructional Model: Origins, Effectiveness, and Applications*. BSCS.

Castoldi, M. (2012). Costruire l’apprendimento: metodologie didattiche a confronto. De Rossi, M, & Petrucco, C. (2013). *La narrazione digitale per l’educazione e la formazione*. (3), 39-49. Carrocci. http://hdl.handle.net/2318/127306

Dekker, S., & Van Baren-Nawrocka, J. (red.) (2017). Wetenschappelijke doorbraken de klas in! Molecuulbotsingen, Stress en Taal der Zintuigen. Nijmegen: Wetenschapsknooppunt Radboud Universiteit.

Ministero dell’Istruzione, dell’Università e della Ricerca (MIUR). (2012). Indicazioni nazionali per il curricolo della scuola dell’infanzia [num. spec.]. *Annali della pubblica istruzione, LXXXVIII.*

Dewey, J. (1910). Science as Subject-Matter and as Method. *Science*, *31*(787), 121–127. https://www.jstor.org/stable/1634781. doi:10.1126cience.31.787.121 PMID:17744061

Duran, L. B., & Duran, E. (2004). The 5E Instructional Model: A Learning Cycle Approach for Inquiry-Based Science Teaching. *The Science Education Review*, *3*(2), 49–58.

Earl, L. M. (2003). Assessment as Learning: Using Classroom Assessment to Maximize Student Learning. Thousand Oaks (CA): Corwin Press European Commission / EACEA / Eurydice, 2022. Increasing achievement and motivation in mathematics and science learning in schools. Eurydice report. Luxembourg: Publications Office of the European Union.

Franco, R. (2018). *L'IBSE: un approccio didattico vincente per l'insegnamento nelle scuole superiori. Il caso innovativo delle scienze della terra «investigate» attraverso l'arte*. Lampi di Stampa.

Harley, W., & Elstgeest, J. (1985). *Primary Science: Taking the Plunge. How to Teach Primary Science More Effectively*. Heinemann.

Lee, V. S., Greene, D. B., Odom, J., Schechter, E., & Slatta, R. W. (2004). What is inquiryguided learning. In V. S. Lee (Ed.), Teaching and learning through inquiry: A guidebook for institutions and instructors (pp. 3-15). Sterling, VA: Stylus Publishing.

Marconato, G., Molinari, M., & Musco, F. (2016). *Competenti si diventa*. Pearson Longman.

Peeters, M. H. J., & Meijer, W. (2014). Onderzoekend leren: Hoe stel je een onderzoeksvraag op? *Jeugd in School en Wereld*, *5*(9), 6–9.

Rocard, M. (2007). *Science Education Now: A Renewed Pedagogy for the Future of Europe*, Rapporto EU22-845, Bruxelles. https://ec.europa.eu/research/sciencesociety/document_library/pdf_06/report-rocard-on-science-education_en.pdf

Rothstein, D., & Santana, L. (2013). *Make just one change: teach students to ask their own questions*. Harvard Education Press.

Spronken-Smith, R. (2008). *Experiencing the Process of Knowledge Creation: The Nature and Use of Inquiry-Based Learning in Higher Education*. Semantic Scholar.

Testoni, A. (2019). *Metodi e contenuti per un corso di chimica nel primo biennio della scuola secondaria di secondo grado. Chimica: Cosa insegnare ai vari livelli scolastici, metodologie didattiche (anche non formali), approccio integrato all'insegnamento*. Società Chimica Italiana.

ENDNOTES

[1]: Programme for International Student Assessment (PISA) conducts international research promoted by The Organisation for Economic Cooperation and Development (OECD). PISA measures 15-year-olds' ability to use their reading, mathematics and science knowledge and skills to meet real-life challenges

[2] : Refer to "Increasing achievement and motivation in mathematics and science learning in schools", a study by Rete Eurydice for further information

[3] : TEMI (Teaching Enquiry with Incorporated Mysteries) is a 42-month European project for the didactic development in sciences, financed by the European Commission FP7 (Social Science)

APPENDIX A

Figure 3. Critical thinking skills

Note taken from https://www.teachthought.com/wp-content/uploads/2013/09/blooms-question-stems.jpg

CRITICAL THINKING SKILLS

Level	Verbs	Question stems
1 Knowledge Identification and recall of information	define, label, name, state fill in the blank, locate, recall, tell list, match, spell, underline identify, memorize	Who ____________? What ____________? Where ____________? When ____________? How ____________? Describe ____________. What is ____________?
2 Comprehension Organization and selection of facts and ideas	convert, interpret, restate, summarize describe, paraphrase, retell in your own words, trace explain, put in order, rewrite, translate	Re-tell ________ in your own words. What is the main idea of ________? What differences exist between ________? Can you write a brief outline?
3 Application Use of facts, rules, and principles	apply, demonstrate, give an example, show compute, determine, illustrate, solve conclude, draw, make, state a rule or principle construct, find out, operate, use	How is ____ an example of ____? How is ____ related to ____? Why is ____ significant? Do you know of another instance where _____? Could this have happened in ____?
4 Analysis Separating a whole into component parts	analyze, contrast, diagram, examine categorize, debate, differentiate, infer classify, deduct, dissect, specify compare, determine the factors, distinguish	What are the parts or features of _______? Classify ______ according to _______. Outline/diagram/web/map __________. How does ____ compare/contrast with ___? What evidence can you present for _____?
5 Synthesis Combining ideas to form a new whole	change, find an unusual way, predict, revise combine, formulate, pretend, suggest compose, generate, produce, suppose construct, invent, rearrange, visualize create, originate, reconstruct, write design, plan, reorganize	What would you predict/infer from _____? What ideas can you add to _____? How would you create/design a new _____? What solutions would you suggest for ____? What might happen if you combined _____ with _____?
6 Evaluation Developing opinions, judgements, or decisions	appraise, decide, judge, rate choose, defend, justify, select compare, evaluate, prioritize, support conclude, give your opinion, rank, value	Do you agree that _____? Explain. What do you think about _____? What is most important? Prioritize _____ according to _____? How would you decide about _____? What criteria would you use to assess ____?

APPENDIX B

Figure 4. The question machine
Note taken from https://www.ru.nl/wetenschapsknooppunt/english/materials/

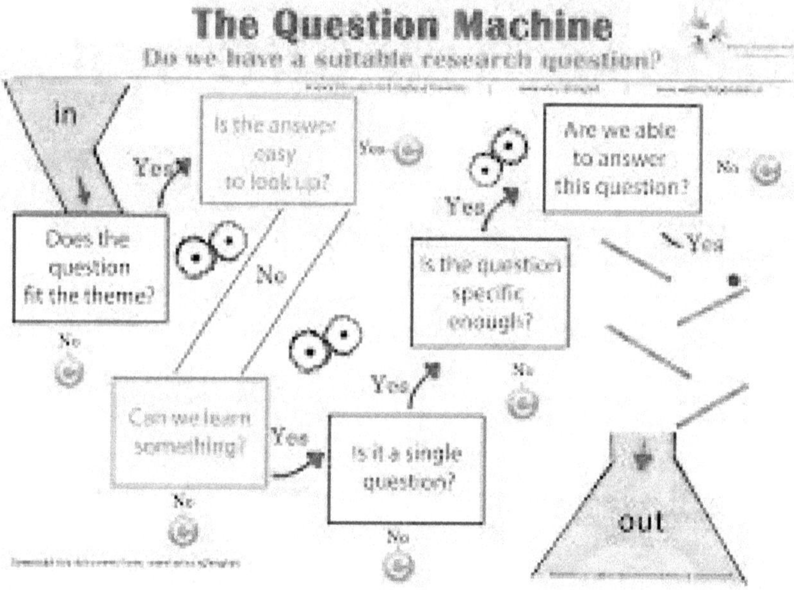

APPENDIX C

Figure 5. Research plan

RESEARCH PLAN

Names:..

1. What is our research question?

....

....

2. What is our hypothesis?

....

....

3. What do we think of doing to check the correctness of the hypothesis?

....

....

....

....

4. What do we expect will happen?

....

....

....

....

5. What material do we need?

- ....
- ....
- ....
- ....
- ....

6. What do we have to observe?

....

....

....

7. How do we note down the results when searching?

....

....

....

....

8. How do we organize ourselves?

WHO?	WHAT?	WHEN?	WHERE?

APPENDIX D

Figure 6. Working plan

WORKING PLAN

Names:..

1. Observations

DATE	MOZZARELLA A	MOZZARELLA B	MOZZARELLA C	MOZZARELLA D

2. What is our conclusion?

....

....

....

....

3. Does the conclusion coincide with our hypotheses? Why?

....

....

....

....

APPENDIX E

Figure 7. Observation check list

OBSERVATION CHECK LIST Name:............................			
	YES	NO	PARTLY
Takes part freely in the conversations			
Takes part in the group work			
Collaborates with classmates			
Helps classmates in trouble			
Asks for help if he is in trouble			
Defends his ideas			

APPENDIX F

Figure 8. Rubric map

RUBRIC – MAP

Criteria	A	B	C	D
Organization	It contains the main concepts in an appropriate number and is well organized and easy to follow	Contains most of the important concepts. It's generally well organized and easy to follow.	Contains only some of the most important concepts. Is organized well enough but sometimes is incoherent	Contains a limited number of important concepts. Is discontinuous and confused.
Concept and terminology	A good understanding and the complexity of thought about the topic is evident: appropriate terminology is used	The thought about the topic is effectively represented; there are some conceptual misunderstandings and some mistakes in terminology	Lacks understanding of the most important concepts and the representation of the topic is just sufficient; there are several mistakes in terminology.	A lack of understanding of the subject is highlighted.
Relationship between concepts	Shows comprehension of the relationship between the main concepts: the relative importance of the concepts is well represented; the relationships are precisely defined.	Some connections between the main concepts are wrong; the relative importance of concepts is sufficiently represented; the relationships are poorly defined	There are a lot of wrong connections; the relative importance of concepts is scarcely represented; the relationships are vaguely defined.	There is a complete lack of the important concepts and the appropriate connections; there is no hierarchy of ideas

Note Taken from "Competenti si diventa", di G. Marconato, M. Molinari, F. Musco, Pearson

APPENDIX G

Figure 9. I Self-assessment: I'm able to

I SELF-ASSESSMENT

Name:............................

1. I'm able to:

	YES	PARTLY, BECAUSE	NO, BECAUSE
Formulate investigative questions		....	....
Formulate hypothesis and previsions		....	....
Link the new experience and a well-known experience		....	....
Communicate my ideas		....	....
Provide evidence to support my arguments		....	....

Figure 10. I Self-assessment: I'm in the group

2. "I in the group"

	Yes	sometimes	no
I respected the role assigned			
I finished my part of the job on time			
I finished my part of the job on schedule			
I treated my classmates with respect			
I asked questions if I didn't understand, and I asked for help when I was in trouble			
I listened carefully while the others were talking			
I helped who was in trouble			
I verified that all the members of the group understood			
I helped to solve problems with serenity			

Figure 11. I Self-Assessment: We in the group

3. "We in the group"

	Yes	sometimes	No
we respected the role assigned			
we involved all the members of the group in the tasks			
we created a good studying atmosphere			
we helped whoever was in trouble			
we verified that everybody understood			
we finished our work on time			
we finished our work on schedule			

APPENDIX H

Figure 12. Cognitive autobiography

COGNITIVE AUTOBIOGRAPHY

Name: ....

1. For each stage of my work, I colour the box that represents the way I have worked

	I NEEDED TO HELP	I HAD SOME TROUBLES	I WAS AUTONOMOUS	I WAS AUTONOMOUS AND I BROUGHT MY CONTRIBUTION
ENGAGE				
EXPLORE				
EXPLAINE				
PROCESS				
EVALUATE				

2. My description of the activity that aroused greater interest and curiosity in me.

....

....

....

3. How did I overcome the difficulties I encountered?

....

....

....

4. What I learned and how I learned: I will tell you about it.

....

....

....

Chapter 10
Pre-Service Teachers' Cognition of STEM via 4IR Cognitive Principles

Anass Bayaga
Nelson Mandela University, South Africa

ABSTRACT

In the current Fourth Industrial Revolution (4IR), or Industry 4.0, learning—and consequentially learners' comprehension—either in a discipline-specific (science, technology, engineering, and mathematics- STEM disciplines) or general decision-making, can be marred by inappropriate use of the cognitive tenets, particularly so in the current 4IR or Industry 4.0. Accordingly, the current study is informed by this challenge and in part anchored upon theoretical disposition of 4IR or Industry 4.0 cognitive principles. The present research provides both theoretical and empirical examinations to improve mathematical discipline cognition as well as general decision making, thus aimed to enhance knowledge-building and subsequent development in discipline-specific.

INTRODUCTION AND CONTEXT

Recently, questions have been raised about how preservice teachers are struggling to apply their knowledge. In a decade too, there have been concerns regarding the lack of expertise in preservice teachers' cognition of science, technology, engineering, and mathematics (STEM) through the fourth industrial revolution (4IR) cognitive principles, mainly so is its applicability and implementation (Bal & Doğanay, 2014; Kandemir & Karadeniz, 2021; Stohlmann et al., 2013; Tan & Ang, 2013). For

DOI: 10.4018/978-1-6684-6932-3.ch010

example, Kandemir and Karadeniz (2021) highlight preservice teachers' cognitive and metacognitive processes in integrated STEM modeling activity. As with the work of Kandemir and Karadeniz (2021) and Tan and Ang (2013), Stohlmann et al. (2013) highlight the need to incorporate preservice elementary teachers' mathematical content knowledge in an integrated STEM modelling. On the other hand, Bal and Doğanay (2014) suggest the need to improve primary school prospective teachers' understanding of the mathematical modeling process both in theory and practice.

Thus, the context of STEM cognition via 4IR cognitive principles in the current study emanates from the highlights and challenges. However, in both developed and developing countries such as South Africa, several studies have attempted to contextualize STEM cognition via 4IR cognitive principles without firm results (Brynjolfsson & McAfee, 2014; Chaka, 2020; Gravemeijer, Stephan, Julie, Lin & Ohtani, 2017; Lee, Lee & Wong, 2021; Lewis, 2020; Merritt, Lee, Rillero & Kinach, 2017; Niss & Højgaard, 2019). For instance, Chaka (2020) notes the skills, competencies, and literacies attributed to 4IR/industry 4.0. Equally, Brynjolfsson and McAfee (2014) likened the present era's work, progress, and prosperity to a machine age. Meanwhile, Gravemeijer et al. (2017) question how "...mathematics education may prepare students for the society of the future."

Consequent to the studies mentioned above, several conclusions have been drawn. For example, Lewis (2020) suggests that the pervasive use of technology in the workplace requires redefining skills for the 21st century. Similarly, both Lee et al. (2021) and Merritt et al. (2017) propose an increased problem-based learning approach in K-8 mathematics and science education, coupled with Niss's and Højgaard's (2019) view on mathematical competencies revisitation.

The popularity of STEM started in the early 2000s intending to increase demand for qualified technical workers, strengthen, innovate, develop the country's competitive disposition globally and simultaneously improve economic sustainability. In essence, STEM is an educational program that prepares students for learning institutions and careers in science, technology, engineering, and mathematics. Beyond subject-specific learning, it also aims at fostering another skill set, such as discovering, inquiring, logical, and collaborative reasoning. The skillsets are essential for current global connectivity, advancements in not just analytics, augmented, mixed, virtual reality (AR, MR, and VR), blockchain technology, and automation, but a requirement in robotics, which jointly is referred to as industry 4.0 or the fourth industrial revolution (4IR).

In preservice teacher preparation, several implications are drawn from the skills set of 4IR in terms of preservice teachers' cognitive processes, such as working memory, inhibitory control, and attention. Several studies, including the work of Oberauer (2019) and Raver and Blair (2016), examined cognitive processes known as executive function (EF). A consensus on what working memory refers to is "the mechanisms and processes that hold the mental representations ....and needed for an ongoing cognitive task available for processing" (Oberauer, 2019, p. 2). Meanwhile, Maraver et al. (2016, p. 2) also defined working memory as "a cognitive system for temporarily storing and managing information that is necessary for undertaking complex cognitive tasks, and it is thought to play a key role in guiding goal-oriented behavior [u]r and novel problem-solving." Though there are also distinctly different definitions of attention, as noted by Oberauer (2019, p. 2), one definition of attention characterizes it as "a limited resource for information processing ... or a process of (or mechanism for) selection of information to be processed with priority".

One implication from the cognitive principles is that for working memory, attention, and inhibitory control to work effectively in achieving the skills set and to improve knowledge-building, the processes must "retrieve information selectively so that only information useful for the current skills set enters working memory [for instance]" (Oberauer, 2019, p. 2). Consequently, the need to examine how preservice teachers apply or should apply cognition principles and develop expertise in STEM cognition through 4IR.

Debatably, STEM cognition, knowledge-building, and subsequent development in discipline-specific over time have been influenced by cognitive principles, especially in industry 4.0. These principles (to be explained later) include (1) modelling and simulation, (2) computational algorithms, and (3) abstract and concrete representations of ideas (Core Team, 2019; González-Calero et al., 2019; Hartanto et al., 2018; Mielicki et al., 2017). In the context of the chapter, the principles need to be decomposed in the South African educational context. Such examinations are vital and consequently conditioned upon the fact that one's experiences inside or outside of a South African classroom, as well as prior instruction and parental expectations, do to no small extent differ across South Africa and comparable countries. Consequently, such differentials or influencers affect STEM cognition in different contexts, hence the need for the current research.

Due to the enormous variability that persists in different contexts and practices, the cognitive principles need to be adequately understood and even consistently applied, particularly in developing countries such as South Africa. For example, STEM textbooks in South Africa predominantly include decorative features in the instructional processes alongside STEM images. However, they need more appropriate or contextual situations for studying and explanations (McLaren et al., 2015). South Africa's practices tend to emulate those similar in STEM cognition in parts of the world with traditionally higher STEM achievement. For instance, teachers in Australia, New Zealand, Hong Kong, and Japan place emphasis preferably on the connections between abstract and concrete representations (Richland & Begolli, 2016). Japanese textbooks focus more on work examples (McLaren et al., 2015). Singaporean students tend to focus more on concrete models (Hoven & Garelick,

2007). While it is intuitive to suggest the use or even adopt such practices that have consistently worked in such countries or have outperformed South Africa, it is still prudent to examine their applicability and implementation in knowledge-building and subsequent development in discipline-specific (mathematical disciplines).

Aim of the Study

The current study is thus rested upon knowledge-building coupled with a need for more development of expertise in (1) modelling and simulation, (2) computational algorithms, and (3) abstract and concrete representations of ideas. While the current study examines knowledge-building, it is partly anchored upon the theoretical disposition of cognitive principles enhancement in industry 4.0. The present conceptual research thus provides both theoretical and empirical examinations to improve STEM discipline cognition through modelling and simulation, computational algorithms, and abstract and concrete representations of ideas.

Related Work

Several factors have motivated for the current study. Compared to in-service teachers' exposure to continuous development program, preservice teachers' cognition arguably could be exacerbated by inappropriate preparatory practices and a lack of applying appropriate 4IR (industry 4.0) cognition principles (Booth & Newton, 2012). Several studies have also attempted to compare factors such as cognitive processes, foundational 4IR or industry 4.0 skills, calculation accuracy, fluency in solving word problems, and pre-algebraic knowledge in industry 4.0 (Fuch et al., 2016). Others examined; linking gestures and cross-cultural variation during instructional analogies (Richland & Begolli, 2016), how language influences reversal error (González-Calero et al., 2019), bilingualism, symbolic abstraction, and their ramifications for STEM (Core Team, 2019; Hartanto et al., 2018; Mielicki et al., 2017). For instance, due to the underperformance related to fluency and proficiency in STEM cognition, in conventional classrooms, Falloon's (2015) study encouraged collaborative learning through iPads. While the majority of other researchers speak about various modalities, including 4IR technology (Booth, 2015a; Booth et al., 2015b; Dunlosky et al., 2013; Pashler et al., 2007; Richland & Begolli, 2016), others such as Richland and Begolli (2016) suggest improving STEM cognition is better conducted via comparison – which could be achieved through analogue visibility. There is even a view of an ecological approach to visual perception, as suggested by Gibson (2014). Whereas Weinberg, Dresen and Slater (2016) suggest students' cognition, for instance, in algebraic notation through a semiotic system, Belenky and Schalk (2014), on the other hand, emphasized learning through transfer and interest. Wherein such a learning process involves an organization of a framework for categorizing external knowledge representations as well as examples in real-world (Booth et al., 2015b; Falloon, 2015; Fleischman et al., 2010; Gibson, 2014; Mullis et al., 2012; Newman, 2016; Wasner et al., 2016). There is also a call for incorrect models to enhance learning (Durkin & Rittle-Johnson, 2012). Nevertheless, Cummins (1992) suggests the need to improve cognition via analogical reasoning during the induction of problem-solving classifications and to develop higher cognitive processes (Vygotsky, 1978).

The conclusion is that preservice teachers' cognition in STEM fluency and proficiency, while not a new trend in Europe, America, and South Africa, but difficult to comprehend and persistently endures, particularly in computation, STEM modelling, and simulation within industry 4.0.

Computational Algorithms

Active learning in STEM arguably occurs when preservice teachers must explain the algorithms in science or mathematical concepts (Booth et al., 2015b). Several reasons have been attributed to this form of learning. One reason invokes the notion that algorithms tend to reduce preservice teachers' cognitive load, which is a consequence of the reduction of both attention and working memory required for all steps involved in solving problems. The principle of algorithm thus requires a focus of the working memory on comprehending thought processes underpinning procedural steps and knowledge as opposed to factual knowledge. Inherent in this process is that computational algorithms give preservice teachers self-explanatory opportunities. It also includes ensuring that expertise is explicit rather than implicit while simultaneously linking new information with prior experience and generating sufficient inferences to fill the gaps in knowledge (Booth et al., 2015b).

Evidence from Laboratory Studies

Within the domain of STEM, algorithm principles have been applied in laboratory research. The evidence demonstrates that algorithms increase problem-solving ability, for instance, in learning probability, algebra, geometry business math (Hsiao et al., 2013). Nevertheless, two scenarios stand out; the benefits derived from such do not require split attention to two separate sources of information. Additionally, it is proven that it decreases the time preservice teachers take to learn mathematical concepts.

Evidence From Classroom Studies

For the most part, studies that have taken place have solely relied on real-world and computer-assisted instruction. For instance, such studies have concluded a benefit in the time required for preservice teachers to learn geometry (Tran et al., 2017; Schwonke et al., 2007). Other evidence suggests increased procedural performance in fractions amongst 5th graders (Lee & Chen, 2015). For secondary school preservice teachers learning engineering and mathematics concepts via algorithm-based podcasts, there was an increase in performance and attitudes (Kay & Edwards, 2012). Evidence has also been linked with a conceptual understanding of algebra learning and explaining algebra in computer-enabled environments (Booth et al., 2013). The assessment is based on the growing body of research in a predominately traditional environment. For instance, demonstrable evidence suggests an inverse relationship between the study of algorithms in algebra and factors such as; fewer problem-solving errors and time to learn, which could be with less assistance from inspectors (Kay & Edwards, 2012).

Recommendations for Further Research

Evidence of the algorithm's effectiveness has been tested mainly in a computerized and real-world situation. However, the evidence demonstrates that even though the algorithm benefits, some exceptions suggest that the algorithm does not work. (Booth et al., 2015b). For instance, evidence is needed to verify prevailing or enabling conditions for algorithm mechanisms. For example, too, are the prevailing or enabling conditions hinged upon varying mathematical contents? Furthermore, how will comprehensive transitional modelling and simulation work benefit real-world situations?

Currently, the majority of studies tend to focus instead on the relationship between prior knowledge and its influence on topics students tend to learn. However, until now, few studies have established the link between prior knowledge regarding different mathematics topics. For instance, what prerequisite skills do students require to learn effectively about some topics? Furthermore, because of such questions, another question is whether learning algorithms will improve that prior knowledge. In essence, further research needs to be conducted to establish such questions.

Summary

In summary, it is fundamental to note that computational algorithms do not work in all cases. Therefore, transitional modelling and simulation beneficial to real-world situations must be adapted to sustain human-to-machine interactions.

STEM Cognition Through Modelling and Simulation in Industry 4.0

The principle of STEM modelling and simulation is mainly derived from the direction of scaffolding, which is eminent from Vygotsky's (1978) zone of proximal development (ZPD), which is sequential. That is, starting from no modelling and simulation, followed by fixed modelling and simulation, and then variable modelling and simulation. In simple terms, in the current industry 4.0, where there may be no modelling and simulation, preservice teachers are encouraged to proceed with fixed modelling and simulation and then variable modelling and simulation. Potential development is supported by an agent (STEM modelling and simulation), in this case, modelling or simulation. Accordingly, the assumption is that preservice teachers' degree of cognition is contingent upon problem-solving, akin to the skills of 4IR (Kupers, van Dijk & van Geert, 2015).

Evidence From Laboratory Studies in Respect of STEM Cognition

Evidence from laboratory studies demonstrates that the degree of modelling or simulation is conditioned upon; the preservice teacher's prior knowledge about the content, regardless of whether it is advanced or elementary science (Schwonke et al., 2011). The assertion emanates from the assessment of Schwonke et al. (2011) and Belland et al. (2015), who compared fixed modelling and simulation to those contingents on students' ZPD. The conclusion was that unlike modelling and simulation, which are contingent on students' ZPD, in contrast, predetermined modelling and simulation delayed the transition of the cognitive process (Belland et al., 2015). Therefore, providing no modeling or simulation may be more effective than delivering predetermined levels of scaffolding to all participants.

Evidence From Classroom Studies in Industry 4.0

The recommendation is that modelling and simulation are essential in the early stages of learning (Smit et al., 2013). Nevertheless, modelling and simulation should be removed in succession to enable a smooth transition. The implication is that the delay in the removal of modelling and simulation and the need for more attention to students' competence levels may lead to underperformance. As with laboratory evidence, research from the classroom also demonstrates that modelling and simulation success in STEM cognition is contingent upon several factors. These include individualized instruction, which must be tailored through 4IR technologies or adaptive learning technologies strategies instead of a size fits all modelling and simulation approach (Schwonke et al., 2011).

Recommendations for Further Research

Evaluation by Schwonke et al. (2011) suggest that modelling and simulation require a more advanced and competent instructor to assist the preservice teacher in solving problems beyond their cognitive abilities. Thus, intelligent tutoring systems in current 4IR need to use the preservice teacher's prior knowledge to determine the degree of the succession of removal of modelling or simulation needed for individualized learning. Hence, making it human-to-machine interaction as opposed to human-to-human communication.

Summary

Based on the recommendation, modelling and simulation require evolving competent preservice teachers with advanced 4IR skill sets, paving the way for human-to-machine interaction instead of human-to-human communication. The crucial thing, though, is to model or simulate a phenomenon or phenomenon of interest for preservice teachers, followed by gradual removal of the modelling or simulation. The withdrawal should be over time as the preservice teachers' conception and procedural knowledge advance. Because of the first, the second requires that modelling and simulation be contingent upon preservice teachers' level of cognition.

Abstract and Concrete Representations of Mathematical Principles

Whether more benefit is derived from students learning from a concrete representation instead of abstract illustration, both debates continue to endure. Even though earlier studies did pitch both concrete and conceptual representation of concepts against each other, a significant body of knowledge suggests that (1) each has its purpose (2) as opposed to pitching the reported advances over the other. It is suggested that a synergistic benefit is only achieved when both concrete and abstract representations are combined (De Bock et al., 2011). For instance, demonstrable evidence suggests that two-digit arithmetic, learned by seven years pupils, was enabled via the linkage between concrete and symbolic representations, wherein the written representations assisted in achieving proper conceptual understanding and transferable knowledge (Uttal et al., 2013). In other studies, Richland and McDonough (2010) were able to reveal the relationship between permutation and combination (probability) by gesturing back and forth between the differences and similarities of the two problems (Kaminski et al., 2013).

Evidence From Laboratory Studies

A preservice teacher effectively learns by connecting a concept's concrete representation and an abstract illustration (Pashler et al., 2007). That is because, while tangible representations are generally considered to be grounded with new information considering prior knowledge, in contrast, abstract representations do assist students in transforming or transferring the unique ability to related use cases (see Belenky & Shalk, 2014).

Evidence From Classroom Studies

Besides linking both representations, gradually fading concreteness with increasingly substituting it with abstract representations may aid efficient cognitions (Braithwaite & Goldstone, 2015). The evidence is that for more complex problems, abstract representations are rather beneficial to McNeil and Uttal (2009) among primary school preservice teachers and undergraduates (Booth, 2015a). On the other hand, preservice teachers s associated with poor prior knowledge instead of benefiting more from concrete, well-grounded representations (Braithwaite & Goldstone, 2015; Petersen & McNeil, 2013). Meanwhile, older preservice teachers benefit more from abstract representations (Booth et al., 2015b; Braithwaite & Goldstone, 2013; McNeil & Fyfe, 2012; Stephan & Akyuz, 2012). The consensus is that appropriateness and effectiveness of concrete and abstract representations tend to vary with preservice teacher features such as age and prior knowledge, which are sometimes confounded.

Recommendations for Further Research

A recommendation from the aforementioned is that given different learning goals and diverse preservice teachers, there is a need to establish which type, combination, and ordering of the most significant representations. There is also a need to further research further on precise and consistent forms of representations, learning goals, and preservice teacher characteristics. Since representations, the tasks and learning goals tend to be confounded and affect both concrete and abstract representation. There is a need for more generalizable evidence that informs practitioners to choose the best sequence of representations for students and classroom objectives.

Summary

In summary, abstract and concrete representations of STEM principles need generalizable evidence to inform practitioners to choose the best sequence of representations for students and classroom objectives. Simultaneously, there is also a need for different learning goals due to diverse preservice teachers.

DISCUSSION

Even though the reviewed principles tend to have brief relevance in STEM cognition with a significant impetus to improve our cognition in STEM computation, each principle varies remarkably (Mullis, 2012). For instance, collectively, Fuch et al. (2016) and Dunlosky et al. (2013) reported that these differences substantially range from quantity to the quality of established studies supporting their degree of effectiveness. For instance, while some of the principles have received testable reviews from laboratory research, there is, however, a need to develop the standard of when and how effective there are for given educational settings (Begolli & Richland, 2016). For instance, as reported by Begolli and Richland (2016), it is believed that while there is increasing evidence that supports computational algorithms, insufficient research details various ways that computational algorithms could and should be used. Such assessment is significant because, for instance, for modelling and simulation, it is rather vital to pay attention to implementation as well as testing in educational settings where there needs to be more assistance for individualized technology. Thus far, the evidence sufficiently supports laboratory and classroom research, particularly in abstract/concrete representations (Belland et al., 2015; Kupers et al., 2015). That is, while learning complex content tends to require abstract representations, straightforward content is more effective with concrete representations (Booth et al., 2013; Kay & Edwards, 2012; Lee & Chen, 2015; Dumas et al., 2016). For illustration, worked examples tend to relate more to straightforward content, like basic science skills, and complex content, such as probability and algebra (Hsiao et al., 2013; Smit et al., 2013).

Similarly, modelling and simulation enhance early years' learning of simple and complex content such as feedback, analogical comparison, and error analysis. However, with and without technology, evidence suggests that implementing the principles has proven to be difficult or even vary markedly (Booth et al., 2013; Kay & Edwards, 2012; Lee & Chen, 2015). Additionally, the nature of the principles in traditional classrooms and the question of when to provide individualized computer programs also need to be clarified (Vamvakoussi & Vosniadou, 2012). Finally, whether they may be incorporated in written materials relations unclear, or whether teachers should actively implement them remains to be established.

A general observation this far is to conduct translational research. Thus, ensuring actual research results reach the needful and equally improve educational settings (Braithwaite & Goldstone, 2013). While disseminating and speculating about research results related to STEM cognition and its associated disciplines, such as cognitive principles, do to some extent expand our understanding of scientific cognition in general, this rarely has a significant impact in a real-life situation (Braithwaite & Goldstone, 2015; Stephan et al., 2015). For example, a limited number of practitioners access research outputs in academic journals and must be regularly exposed to the current state of affairs. Additionally, the notion that some findings in the laboratory research work do not fundamentally suggest the same is true for classroom research. This is because there could be a mismatch or failure to translate to the classroom, hence appealing to only a limited proportion of those it was intended for (D'Mello & Graesser, 2014; Overoye & Storm, 2015). Because most classes are less controlled than university laboratories, findings related to translating laboratory research to classroom do, to a more considerable degree, depend on some factors—an innate understanding of the relationship between both principles and education setting. It is also essential to consider developing interactive and iterative lesson materials. The motive is to possess diverse opportunities for revision during field testing and experimental design. This Translational research works best when collaborating with teachers and administrators. Recognising that a workable instruction should not fit all is also essential. Factors such as individual differences that consider sub-factors such as prior knowledge and working memory capacity are essential for future research. Particularly true are demographic factors such as gender and race, which may significantly affect the interpretation of cognitive principles and how they influence instructional materials to elevate the benefits, particularly for the actual classroom situation.

CONCLUSION AND CONTRIBUTION OF THE CURRENT STUDY

Based on these cognitive principles, the current research examined the degree of agreement concerning STEM cognition and knowledge-building. First, based on the motivations, we extend our understanding of the evidence for principles important for improving STEM cognition. We achieve this by examining how to improve cognitive knowledge via, for example, STEM modelling and simulation using adaptive e-learning technologies, computational algorithms, and abstract and concrete representations of STEM cognitive principles. Accordingly, the present work examines laboratory and classroom research in various ways. Second, conditioned upon such examinations, attention was extended to a summary of; known and unknown ideas emanating from the study and how they are linked with mathematics cognition, followed by gaps in research. Third, an examination of the present state of facts relating to how such principles influence both elementary and advanced mathematics conceptualization is assessed. Forth, we simultaneously develop recommendations worthy of impacting the regulations regarding mathematics cognition and knowledge–building.

FUTURE RESEARCH

Two critical future works are recommended. First, there is a need to conduct translational research related to the principles to reach end-users and equally improve educational settings. Conducting research is vital because, while speculating expands our understanding of scientific cognition, it rarely significantly impacts a real-life situation. Second, while a gradual removal of modelling and simulation is thus encouraged, nevertheless, there is still a need for research on how and when it is appropriate for such gradual withdrawal for different mathematical cognition. While fading of support is necessary, further work should investigate how and when it is best to fade support for additional accurate content, which may include cognitive characteristics such as working memory, inhibitory control, and attention. The reason is that they impede learning when modelling and simulation are inadequate or exceedingly higher. Additionally, due to challenges associated with access, reliability, and ease of use of computer-based learning, it will be prudent to research methods.

REFERENCES

Bal, A. P., & Doğanay, A. (2014). Improving primary school prospective teachers' understanding of the mathematical modeling process. *Educational Sciences: Theory & Practice*, *14*(4), 1375–1384.

Belenky, D. M., & Schalk, L. (2014). The effects of idealised and grounded materials on learning, transfer, and interest: An organising framework for categorising external knowledge representations. *Educational Psychology Review*, *26*(1), 27–50. doi:10.100710648-014-9251-9

Belland, B. R., Walker, A. E., Olsen, M. W., & Leary, H. (2015). A pilot meta-analysis of computer-based scaffolding in STEM education. *Journal of Educational Technology & Society*, *18*(1), 183–197.

Booth, J. L., Cooper, L. A., Donovan, M. S., Huyghe, A., Koedinger, K. R., & Paré-Blagoev, E. J. (2015a). Design-based research within the constraints of practice: Algebra by Example. *Journal of Education for Students Placed at Risk*, *20*(1-2), 79–100. doi:10.1080/10824669.2014.986674

Booth, J. L., Oyer, M. H., Paré-Blagoev, E. J., Elliot, A., Barbieri, C., Augustine, A. A., & Koedinger, K. R. (2015b). Learning algebra, by example, in real-world classrooms. *Journal of Research on Educational Effectiveness*, *8*(4), 530–551. doi :10.1080/19345747.2015.1055636

Braithwaite, D. W., & Goldstone, R. L. (2015). Effects of variation and prior knowledge of abstract concept learning. *Cognition and Instruction*, *33*(3), 226–256. doi:10.1080/07370008.2015.1067215

Brynjolfsson, E., & McAfee, A. (2014). *The second machine age: Work, progress, and prosperity in a time of brilliant technologies*. W. W. Norton & Company.

Chaka, C. (2020). Skills, competencies, and literacies attributed to 4IR/Industry 4.0: Scoping review. *IFLA Journal*, *46*(4), 369–399. doi:10.1177/0340035219896376

Core Team. (2019). *R: A language and environment for statistical computing*. Vienna, Austria: R Foundation for Statistical Computing." https://www.R- project.org/

Cummins, D. D. (1992). Role of analogical reasoning in the induction of problem categories. *Journal of Experimental Psychology. Learning, Memory, and Cognition*, *18*(5), 1103–1124. doi:10.1037/0278-7393.18.5.1103

D'Mello, S., & Graesser, A. (2014). Confusion and its dynamics during device comprehension with breakdown scenarios. *Acta Psychologica*, *151*, 106–116. doi:10.1016/j.actpsy.2014.06.005 PMID:24973629

De Bock, D., Deprez, J., Van Dooren, W., Roelens, M., & Verschaffel, L. (2011). Abstract or concrete examples in learning mathematics? A replication and elaboration of Kaminski, Sloutsky, and Heckler's study. *Journal for Research in Mathematics Education*, *42*(2), 109–126. doi:10.5951/jresematheduc.42.2.0109

Durkin, K., & Rittle-Johnson, B. (2012). *The effectiveness of using incorrect examples to support learning about decimal magnitude*. Learning and Instruction. doi:10.1016/j.learninstruc.2011.11.001

Falloon, G. (2015). What is the difference? Learning collaboratively using iPads in conventional classrooms. *Computers & Education, 84*, 62–77.

Fuchs, S., Gilbert, J. K., Powell, S. R., Cirino, P. T., Fuchs, D., Hamlett, C. L., Seethaler, P. M., & Tolar, T. D. (2016). The Role of Cognitive Processes, Foundational Math Skill, and Calculation Accuracy and Fluency in Word-Problem Solving versus Pre-Algebraic Knowledge. *Developmental Psychology*, *52*(12), 2085–2098. doi:10.1037/dev0000227 PMID:27786534

Fyfe, E. R., DeCaro, M. S., & Rittle-Johnson, B. (2015). When feedback is cognitive-demanding: The importance of working memory capacity. *Instructional Science*, *43*(1), 73–91. doi:10.100711251-014-9323-8

Gibson, J. J. (2014). The ecological approach to visual perception: classic edition. Psychology Press.

González-Calero, J. A., Berciano, A., & Arnau, D. (2019). The role of language on the reversal error. A study with bilingual Basque-Spanish students. *Mathematical Thinking and Learning*, 1–19.

Gravemeijer, K., Stephan, M., Julie, C., Lin, F., & Ohtani, M. (2017). What mathematics education may prepare students for the society of the future? *International Journal of Science and Mathematics Education*, *15*(S1), 105–123. doi:10.100710763-017-9814-6

Grossnickle, E. M., Dumas, D., Alexander, P. A., & Baggetta, P. (2016). Individual differences in the process of relational reasoning. *Learning and Instruction*, *42*, 141–159. doi:10.1016/j.learninstruc.2016.01.013

Hartanto, A., Yang, H., & Yang, S. (2018). Bilingualism positively predicts mathematical competence: Evidence from two large-scale studies. *Learning and Individual Differences*, *61*, 216–227. doi:10.1016/j.lindif.2017.12.007

Hoven, J., & Garelick, B. (2007). Singapore Math: Using the bar model approach, Singapore textbooks enable students to solve complex math problems—and learn how to think symbolically. *Educational Leadership*, *65*, 38–21.

Kaminski, J. A., & Sloutsky, V. M. (2013). Extraneous perceptual information interferes with children's acquisition of mathematical knowledge—. *Journal of Educational Psychology*, *105*(2), 351–363. doi:10.1037/a0031040

Kandemir, M. A., & Karadeniz, I. (2021). Preservice teachers' cognitive and metacognitive processes in integrated STEM modeling activity. [JESEH]. *Journal of Education in Science, Environment and Health*, *7*(2), 104–127. doi:10.21891/jeseh.832574

Kay, R., & Edwards, J. (2012). A formative analysis of the use of worked example video podcasts in middle school mathematics classrooms. *Canadian Journal of Learning and Technology*, *38*(3). doi:10.21432/T2PK5Z

Kupers, E., van Dijk, M., & van Geert, P. (2015). Within-teacher differences in one-to-one teacher-student interactions in instrumental music lessons. *Learning and Individual Differences*, *37*, 283–289. doi:10.1016/j.lindif.2014.11.012

Lee, C., & Chen, M. (2015). Effects of worked examples using manipulatives on fifth graders' learning performance and attitude toward mathematics. *Journal of Educational Technology & Society, 18*(1), 264-275.40

Lee, N. H., Lee, J., & Wong, Z. Y. (2021). Preparing Students for the Fourth Industrial Revolution through Mathematical Learning: The Constructivist Learning Design. *Korean J. Opt. Photon, 31*(3), 321–356. doi:10.29275/jerm.2021.31.3.321

Lewis, K. (2020). Technology in the workplace: Redefining skills for the 21st century. *The Midwest Quarterly, 61*(3), 348–355.

Maraver, M. J., Bajo, M. T., & Gomez-Ariza, C. J. (2016). Training on Working Memory and Inhibitory Control in Young Adults. *Frontiers in Human Neuroscience, 10*, 1–18. doi:10.3389/fnhum.2016.00588 PMID:27917117

McLaren, B. M., Adams, D. M., & Mayer, R. E. (2015). Delayed learning effects with erroneous examples: A study of learning decimals with a web-based tutor. *International Journal of Artificial Intelligence in Education, 25*(4), 520–542. doi:10.100740593-015-0064-x

Merritt, J., Lee, M. Y., Rillero, P., & Kinach, B. M. (2017). Problem-based learning in K-8 mathematics and science education: A literature review. *The Interdisciplinary Journal of Problem-Based Learning, 11*(2), 3. doi:10.7771/1541-5015.1674

Mielicki, M. K., Kacinik, N. A., & Wiley, J. (2017). Bilingualism and symbolic abstraction: Implications for algebra learning. *Learning and Instruction, 49*, 242–250. doi:10.1016/j.learninstruc.2017.03.002

Mullis, I. V., Martin, M. O., Foy, P., & Arora, A. (2012). TIMSS 2011 international results in mathematics. International Association for the Evaluation of Educational Achievement.

Newman, S. D. (2016). *Does finger sense predict addition performance?* Cognitive Psychology. doi:10.100710339-016-0756-7

Niss, M., & Højgaard, T. (2019). Mathematical competencies revisited. *Educational Studies in Mathematics, 102*(1), 9–28. doi:10.100710649-019-09903-9

Oberauer, K. (2019). Working Memory and Attention – A Conceptual Analysis and Review. *Journal of Cognition, 2*(1), 1–23. doi:10.5334/joc.58

Overoye, A. L., & Storm, B. C. (2015). Harnessing the power of uncertainty to enhance learning. *Translational Issues in Psychological Science, 1*(2), 140–148. doi:10.1037/tps0000022

Pashler, H., Bain, P., Bottge, B., Graesser, A., Koedinger, K., McDaniel, M., & Metcalfe, J. (2007). *Organising instruction and study to improve student learning: IES practice guide (NCER 2007-2004)*. National Center for Education Research.

Piaget, J. (1980). *Experiments in contradictions*. University of Chicago Press.

Raver, C., & Blair, C. (2016). Neuroscientific Insights: Attention, Working Memory, and Inhibitory Control. *The Future of Children*, *26*(2), 95–118. doi:10.1353/foc.2016.0014

Richland, L. E., & Begolli, K. (2016). Teaching Mathematics by Comparison: Analog Visibility as a Double-Edged Sword. *Journal of Educational Psychology*, *108*(2), 194–213. doi:10.1037/edu0000056

Richland, L. E., & McDonough, I. M. (2010). Learning by analogy: Discriminating between potential analogues. *Contemporary Educational Psychology*, *35*, 28–43.

Rittle-Johnson, B., Star, J. R., & Durkin, K. (2012). Developing procedural flexibility: Are novices prepared to learn from comparing procedures? *The British Journal of Educational Psychology*, *8*(3), 436–455. doi:10.1111/j.2044-8279.2011.02037.x PMID:22881048

Schwonke, R., Renkl, A., Salden, R., & Aleven, V. (2011). Effects of different ratios of worked solution steps and problem-solving opportunities on cognitive load and learning outcomes. *Computers in Human Behavior*, *27*(1), 58–62. doi:10.1016/j.chb.2010.03.037

Stephan, M., & Akyuz, D. (2012). A proposed instructional theory for integer addition and subtraction. *Journal for Research in Mathematics Education*, *43*(4), 428–464. doi:10.5951/jresematheduc.43.4.0428

Stohlmann, M. S., Moore, T. J., & Cramer, K. (2013). Preservice elementary teachers' mathematical content knowledge from an integrated STEM modelling activity. *Journal of Mathematical Modelling and Application*, *1*(8), 18–31.

Tan, L. S., & Ang, K. C. (2013). Preservice secondary school teachers' knowledge in the mathematical modeling case study. In *Teaching Mathematical Modelling: Connecting to Research and Practice* (pp. 373–383). Springer. doi:10.1007/978-94-007-6540-5_31

Tran, C., Smith, B., & Buschkuehl, M. (2017). Support for mathematical thinking through embodied cognition: Nondigital and digital approaches. *Cognitive Research: Principles and Implications*, *2*(1), 16. doi:10.118641235-017-0053-8 PMID:28280771

Uttal, D. H., Amaya, M., del Rosario Maita, M., Hand, L. L., Cohen, C. A., O'Doherty, K., & DeLoache, J. S. (2013). It works both ways: Transfer difficulties between manipulatives and written subtraction solutions. *Child Development Research, 2013*, 216–367. doi:10.1155/2013/216367

Vamvakoussi, X., & Vosniadou, S. (2012). Bridging the gap between the dense and the discrete: The number line and the "rubber line" bridging analogy. *Mathematical Thinking and Learning, 14*(4), 265–284. doi:10.1080/10986065.2012.717378

Vygotsky, L. S. (1978). *Mind in society: The development of higher psychological processes*. Harvard University Press.

Wasner, M., Nuerk, H.-C., Martignon, L., Roesch, S., & Moeller, K. (2016). Finger gnosis predicts a unique but small part of the variance in initial arithmetic performance. *Journal of Experimental Child Psychology, 146*, 1–16. doi:10.1016/j.jecp.2016.01.006 PMID:26895483

Weinberg, A., Dresen, J., & Slater, T. (2016). Students' understanding of algebraic notation: A semiotic systems perspective. *The Journal of Mathematical Behavior, 43*, 70–88. doi:10.1016/j.jmathb.2016.06.001

Compilation of References

Abbas, J. R., Kenth, J. J., & Bruce, I. A. (2020). The role of virtual reality in the changing landscape of surgical training. *The Journal of Laryngology and Otology, 134*(10), 863–866. doi:10.1017/S0022215120002078 PMID:33032666

Acharya, S., Bhatt, A. N., Chakrabarti, A., Delhi, V. S., Diehl, J. C., van Andel, E., & Subra, R. (2021). Problem-Based Learning (PBL) in Undergraduate Education: Design Thinking to Redesign Courses. In *Design for Tomorrow* (pp. 349–360). Springer. doi:10.1007/978-981-16-0119-4_28

Adah Miller, E., Codere, S., Charara, J., Severson, A., & Spaul, H. (2022). Fire Ecology in Fourth Grade: Moving Place-based Education to the Global Stage. Suleiman, M., & Huber, T. (eds.) Beyond Provincialism: Promoting Global Competencies in Teacher and Educator Preparation. Information Age Publishing.

Adah Miller, E., Makori, H., Akgun, S., Miller, C., Li, T., & Codere, S. (2022). Including teachers in the social justice equation of project-based learning: A response to Lee & Grapin. *Journal of Research in Science Teaching, 59*(9), 1726–1732. doi:10.1002/tea.21805

Adefila, A., Opie, J., Ball, S., & Bluteau, P. (2020). Students' engagement and learning experiences using virtual patient simulation in a computer supported collaborative learning environment. *Innovations in Education and Teaching International, 57*(1), 50–61.

Adesoji, F. A., Omilani, N. A., & Dada, S. O. (2017). A comparison of perceived and actual; Students' learning difficulties in physical chemistry. *International Journal of Brain and Cognitive Sciences, 6*(1), 1–8.

Aebersold, M., & Dunbar, D. M. (2021). Virtual and Augmented Realities in Nursing Education: State of the Science. *Annual Review of Nursing Research, 39*(1), 225–242. doi:10.1891/0739-6686.39.225 PMID:33431644

Agha, S. (2021). Aligning continuing professional development (CPD) with quality assurance (QA): A perspective of healthcare leadership. *Quality & Quantity*, 1–15.

Aguiar, O. G., Mortimer, E. F., & Scott, P. (2009). Learning from and responding to students' questions: The authoritative and dialogic tension. *Journal of Research in Science Teaching, 47*(2), 174–193. doi:10.1002/tea.20315

Alexander, R. (2006). *Towards dialogic teaching* (3rd ed.). Dialogos.

Alhalabi, W. (2016). Virtual reality systems enhance students' achievements in engineering education. *Behaviour & Information Technology, 35*(11), 919–925. doi:10.1080/0144929X.2016.1212931

Alnagrat, A. J. A., Ismail, R. C., & Idrus, S. Z. S. (2021, May). Extended Reality (XR) in Virtual Laboratories: A Review of Challenges and Future Training Directions. [). IOP Publishing.]. *Journal of Physics: Conference Series, 1874*(1), 012031. doi:10.1088/1742-6596/1874/1/012031

Al-Nuaimy, E. A., Twaij, A. H., & Mahmood, H. A. (2021). Importance of Distance Education and its Rise during the Period COVID-19. *International Journal of Educational Research, 12*(2), 29–34.

Alves, P., Miranda, L., & Morais, C. (2017). The influence of virtual learning environments in students' performance. *Universal Journal of Educational Research, 5*(3), 517–527. doi:10.13189/ujer.2017.050325

Anderson, J., & Li, Y. (Eds.). (2020). *Integrated approaches to STEM education: An international perspective*. Springer Nature. doi:10.1007/978-3-030-52229-2

Annala, J., Lindén, J., Mäkinen, M., & Henriksson, J. (2021). Understanding academic agency in curriculum change in higher education. *Teaching in Higher Education*, 1–18. doi:10.1080/13562517.2021.1881772

Antoniou, P., Arfaras, G., Pandria, N., Ntakakis, G., Bambatsikos, E., & Athanasiou, A. (2020). Real-time affective measurements in medical education, using virtual and mixed reality. In *International Conference on Brain Function Assessment in Learning* (pp. 87-95). Springer, Cham. 10.1007/978-3-030-60735-7_9

Argyris, C. (1991). Teaching smart people how to learn. *Harvard Business Review, 69*(3).

Asim, S., Ponners, P. J., Bartlett, C., Parker, A., & Star, R. (2020). Differentiating Instruction: For Middle School Students in Virtual Learning Environments. *Delta Kappa Gamma Bulletin, 86*(3), 19–31.

Badreddine, Z., & Buty, C. (2011). Discursive reconstruction of the scientific story in a teaching sequence. *International Journal of Science Education, 33*(6), 773–795. doi:10.1080/09500693.2010.496475

Bakhtin, M. M. (1986). *The dialogic imagination: Four essays (M. Holquist* (C. Emerson & M. Holquist Eds. & Trans.). University of Texas Press.

Bal, A. P., & Doğanay, A. (2014). Improving primary school prospective teachers' understanding of the mathematical modeling process. *Educational Sciences: Theory & Practice, 14*(4), 1375–1384.

Balanskat, A., Engelhardt, K., & Licht, A. H. (2018). *Strategies to include computational thinking in school curricula in Norway and Sweden- European Schoolnet's 2018 Study Visit*. European Schoolnet.

Bansal, G. & Ramnarain, U. (2021). Inquiry-Based Science Education in Primary Schools. *Education, 49* (3), 259-262. doi:10.1080/03004279.2020.1854955

Bansal, G. (2018). Teacher discursive moves: Conceptualising a schema of dialogic discourse in science classrooms. *International Journal of Science Education, 40*(15), 1891–1912. doi:10.1080/09500693.2018.1514543

Barreto, L. P., Rodrigues, A. A. D., Oliveira, G. C. B., Almeida, L. T. G., Felix, M. A. C., Silva, P. S., Quadros, A. L., Macedo, A. M., & Mortimer, E. F. (2021). The use of different translation devices to analyze knowledge-building in a university chemistry classroom. *Research in Science Education, 51*(1), 135–152. doi:10.100711165-020-09969-z

Belenky, D. M., & Schalk, L. (2014). The effects of idealised and grounded materials on learning, transfer, and interest: An organising framework for categorising external knowledge representations. *Educational Psychology Review, 26*(1), 27–50. doi:10.100710648-014-9251-9

Belland, B. R., Walker, A. E., Olsen, M. W., & Leary, H. (2015). A pilot meta-analysis of computer-based scaffolding in STEM education. *Journal of Educational Technology & Society, 18*(1), 183–197.

Beneke, S., & Ostrosky, M. M. (2009). Teachers' Views of the Efficacy of Incorporating the Project Approach into Classroom Practice with Diverse Learners. *Early Childhood Research & Practice, 11*(1), n1.

Bereiter, C., & Scardamalia, M. (2014). Knowledge building and knowledge creation: One concept, two hills to climb. In S. C. Tan, H. J. So, & J. Yeo (Eds.), *Knowledge creation in education* (pp. 35–52). Springer. doi:10.1007/978-981-287-047-6_3

Berland, M., & Wilensky, U. (2015). Comparing virtual and physical robotics environments for supporting complex systems and computational thinking. *Journal of Science Education and Technology, 24*(5), 628–647. doi:10.100710956-015-9552-x

Bevins, F., Bryant, J., Krishnan, C., & Law, J. (2020). *Coronavirus: How should US higher education plan for an uncertain future*. McKinsey.

Birt, J., Stromberga, Z., Cowling, M., & Moro, C. (2018). Mobile mixed reality for experiential learning and simulation in medical and health sciences education. *Information (Basel), 9*(2), 31. doi:10.3390/info9020031

Blanchard, N., Donnelly, P. J., Olney, A. M., Samei, B., Ward, B., Sun, X., Kelly, S., Nystrand, M., & D'Mello, S. K. (2016). Semi-automatic detection of teacher questions from human-transcripts of audio in live classrooms. *Proceedings of the 9th International Conference on Educational Data Mining (EDM 2016)* (pp. 288-291). Springer.

Blaschke, L. M. (2012). Heutagogy and lifelong learning: A review of heutagogical practice and self-determined learning. *International Review of Research in Open and Distance Learning, 13*(1), 56–71. https://www.irrodl.org/index.php/irrodl/article/view/1076/2113. doi:10.19173/irrodl.v13i1.1076

Bloom, B. S. (1956). *Taxonomy of educational objectives, handbook: The cognitive domain.* David McKay.

Blumenfeld, P. C., Soloway, E., Marx, R. W., Krajcik, J. S., Guzdial, M., & Palincsar, A. (1991). Motivating project-based learning: Sustaining the doing, supporting the learning. *Educational Psychologist, 26*(3-4), 369–398. doi:10.1080/00461520.1991.9653139

Booth, J. L., Cooper, L. A., Donovan, M. S., Huyghe, A., Koedinger, K. R., & Paré-Blagoev, E. J. (2015a). Design-based research within the constraints of practice: Algebra by Example. *Journal of Education for Students Placed at Risk, 20*(1-2), 79–100. doi:10.1080/10824669.2014.986674

Booth, J. L., Oyer, M. H., Paré-Blagoev, E. J., Elliot, A., Barbieri, C., Augustine, A. A., & Koedinger, K. R. (2015b). Learning algebra, by example, in real-world classrooms. *Journal of Research on Educational Effectiveness, 8*(4), 530–551. doi:10.1080/19345747.2015.1055636

Borko, H., Jacobs, J., Seago, N., & Mangram, C. (2014). Facilitating videobased professional development: Planning and orchestrating productive discussions. In Y. Li, E. A. Silver, & S. Li (Eds.), *Transforming mathematics instruction: Multiple approaches and practices* (pp. 259–281). Springer. doi:10.1007/978-3-319-04993-9_16

Boston Consulting Group. (2022, May 23). *37% of the Top 20 Skills Requested for the Average US Job Have Changed Since 2016*. BCG Global. https://www.bcg.com/press/23may2022-top-20-skills-requested-for-average-us-job

Boulton, C. A., Kent, C., & Williams, H. T. (2018). Virtual learning environment engagement and learning outcomes at a `bricks-and-mortar'university. *Computers & Education, 126*, 129–142. doi:10.1016/j.compedu.2018.06.031

Bower, M., & Falkner, K. (2015). Computational Thinking, the Notional Machine, Pre-service Teachers, and Research Opportunities. In. D'Souza, D and Falkner, K. (Eds.), *Proceedings of the 17th Australasian Computing Education Conference Sydney, Australia*. CRPIT.

Bower, M., Wood, L. N., Lai, J. W., Highfield, K., Veal, J., Howe, C., Lister, R., & Mason, R. (2017). Improving the computational thinking pedagogical capabilities of school teachers. *The Australian Journal of Teacher Education, 42*(3), 53–72. doi:10.14221/ajte.2017v42n3.4

Braithwaite, D. W., & Goldstone, R. L. (2015). Effects of variation and prior knowledge of abstract concept learning. *Cognition and Instruction, 33*(3), 226–256. doi:10.1080/07370008.2015.1067215

Brandon, E., Freiwirth, R., & Hjersman, J. (2021, May). Special Session—Student Engagement with Reduced Bias in a Virtual Classroom Environment. In *2021 7th International Conference of the Immersive Learning Research Network (iLRN)* (pp. 1-3). IEEE.

Bransford, J. D., Brown, A. L., & Cocking, R. R. (2000). *How people learn* (Vol. 11). National academy press.

Brennan, K., & Resnick, M. (2012). *New frameworks for studying and assessing the development of computational thinking*. Paper presented at the Proceedings of the 2012 Annual Meeting of the American Educational Research Association, Vancouver, Canada.

Britain, S., & Liber, O. (1999). *A Framework for Pedagogical Evaluation of Virtual Learning Environments*. Department of Education.

Broadley, K. (2015). Entrenched gendered pathways in science, technology, engineering and mathematics: Engaging girls through collaborative career development. *Australian Journal of Career Development, 24*(1), 27–38. doi:10.1177/1038416214559548

Brynjolfsson, E., & McAfee, A. (2014). *The second machine age: Work, progress, and prosperity in a time of brilliant technologies*. W. W. Norton & Company.

Burde, D., Kapit, A., Wahl, R. L., Guven, O., & Skarpeteig, M. I. (2017). Education in emergencies: A review of theory and research. *Review of Educational Research, 87*(3), 619–658. doi:10.3102/0034654316671594

Burgess, A., van Diggele, C., Roberts, C., & Mellis, C. (2020). Key tips for teaching in the clinical setting. *BMC Medical Education, 20*(2), 1–7. doi:10.118612909-020-02283-2 PMID:33272257

Bybee, R. (1997). *Achieving scientific literacy: From purposes to practices*. Heinemann Publications.

Bybee, R., Taylor, J., Gardner, A., Scotter, P., Carlson, J., Westbrook, A., & Landes, N. (2006). *The BSCS 5E Instructional Model: Origins, Effectiveness, and Applications*. BSCS.

Caballero, D., Araya, R., Kronholm, H., Viiri, J., Mansikkaniemi, A., Lehesvuori, S., Virtanen, T., & Kurimo, M. (2017). ASR in Classroom Today: Automatic Visualization of Conceptual Network in Science Classrooms. In É. Lavoué, H. Drachsler, K. Verbert, J. Broisin, & M. Pérez-Sanagustín (Eds.), *Data Driven Approaches in Digital Education: 12th European Conference on Technology Enhanced Learning, EC-TEL 2017,* (pp. 541-544). Springer International Publishing. 10.1007/978-3-319-66610-5_58

Cabrera, L. (2019). Teacher preconceptions of computational thinking: A systematic literature review. *Journal of Technology and Teacher Education, 27*(3), 305–333.

Castoldi, M. (2012). Costruire l'apprendimento: metodologie didattiche a confronto. De Rossi, M, & Petrucco, C. (2013). *La narrazione digitale per l'educazione e la formazione*. (3), 39-49. Carrocci. http://hdl.handle.net/2318/127306

Chaka, C. (2020). Skills, competencies, and literacies attributed to 4IR/Industry 4.0: Scoping review. *IFLA Journal, 46*(4), 369–399. doi:10.1177/0340035219896376

Chamberlain, J. M., Lancaster, K., Parson, R., & Perkins, K. K. (2014). How guidance affects student engagement with an interactive simulation. *Chemistry Education Research and Practice, 15*(4), 628–638. doi:10.1039/C4RP00009A

Chan, K. K. H., Yu, S. K. K., & Sin, R. K. H. (2021b). Using student actors and video-mediated reflection to promote pre-service teachers' ability to enact responsive teaching. *Innovations in Science Teacher Education. 6*(1). https://innovations.theaste.org/using-student-actors-and-video-mediated-reflection-to-promote-preservice-teachers-ability-to-enact-responsive-teaching/

Chan, K. K. H. (2022). Eliciting and working with student thinking: Preservice science teachers' enactment of core practices when orchestrating collaborative group work. *Journal of Research in Science Teaching, 60*(5), 1014–1052. doi:10.1002/tea.21823

Chan, K. K. H., Xu, L., Cooper, R., Berry, A., & van Driel, J. H. (2021a). Teacher noticing in science education: Do you see what I see? *Studies in Science Education, 57*(1), 1–44. doi:10.1080/03057267.2020.1755803

Chan, K. K. H., & Yau, K. W. (2021). Using video-based interviews to investigate pre-service secondary science teachers' situation-specific skills for informal formative assessment. *International Journal of Science and Mathematics Education, 19*(2), 289–311. doi:10.100710763-020-10056-y

Chan, K. K. H., & Yung, B. H. W. (2015). On-site pedagogical content knowledge development. *International Journal of Science Education, 37*(8), 1246–1278. doi:10.1080/09500693.2015.1033777

Chan, M. C. E., Ochoa, X., & Clarke, D. (2020). Multimodal Learning Analytics in a Laboratory Classroom. In M. Virvou, E. Alepis, G. Tsihrintzis, & L. Jain (Eds.), *Machine Learning Paradigms. Intelligent Systems Reference Library* (Vol. 158). Springer. doi:10.1007/978-3-030-13743-4_8

Chan, S. (2021). *Digitally Enabling'Learning by Doing'in Vocational Education: Enhancing' Learning as Becoming 'Processes.* Springer Nature. doi:10.1007/978-981-16-3405-5

Charara, J., Miller, E. A., & Krajcik, J. (2021). Knowledge in Use: Designing for Play in Kindergarten Science Contexts. *Journal of Leadership, Equity, and Research, 7*(1).

Chin, C. (2007). Teacher questioning in science classrooms: Approaches that stimulate productive thinking. *Journal of Research in Science Teaching, 44*(6), 815–843. doi:10.1002/tea.20171

Chu, S. K. W., Tse, S. K., & Chow, K. (2011). Using collaborative teaching and inquiry project-based learning to help primary school students develop information literacy and information skills. *Library & Information Science Research, 33*(2), 132–143.

Clarke, D., Mitchell, C., & Bowman, P. (2009). Optimising the use of available technology to support international collaborative research in mathematics classrooms. In Janik. T. & T. Seidel (Eds.), The Power of Video Studies in Investigating Teaching and Learning in the Classroom (pp. 39-60). New York: Waxmann.

Clarke-Midura, J., Silvis, D., Shumway, J. F., Lee, V. R., & Kozlowski, J. S. (2021). Developing a kindergarten computational thinking assessment using evidence-centred design: The case of algorithmic thinking. *Computer Science Education, 31*(2), 117–140. doi:10.1080/08993408.2021.1877988

Cobb, P., & Bowers, J. (1999). Cognitive and situated learning perspectives in theory and practice. *Educational Researcher*, *28*(2), 4–15. doi:10.3102/0013189X028002004

Core Team. (2019). *R: A language and environment for statistical computing*. Vienna, Austria: R Foundation for Statistical Computing." https://www.R- project.org/

Corradini, I., Lodi, M., & Nardelli, E. (2017). *Conceptions and misconceptions about computational thinking among Italian primary school teachers*. Proceedings of the 2017 ACM Conference on International Computing Education Research - ICER '17, Tacoma, WA, USA. 10.1145/3105726.3106194

Crouch, L., Rolleston, C., & Gustafsson, M. (2021). Eliminating global learning poverty: The importance of equalities and equity. *International Journal of Educational Development*, *82*, 102250. doi:10.1016/j.ijedudev.2020.102250

Cullen, R. (2002). Supportive teacher talk: The importance of the F-move. *ELT Journal*, *56*(2), 117–126. doi:10.1093/elt/56.2.117

Cummins, D. D. (1992). Role of analogical reasoning in the induction of problem categories. *Journal of Experimental Psychology. Learning, Memory, and Cognition*, *18*(5), 1103–1124. doi:10.1037/0278-7393.18.5.1103

Cunningham, C. M., & Carlsen, W. S. (2014). Unpacking its definitions: A conceptual analysis of integrated STEM education. *Journal of STEM Education: Innovations and Research*, *15*(1), 37–48.

D'Mello, S., & Graesser, A. (2014). Confusion and its dynamics during device comprehension with breakdown scenarios. *Acta Psychologica*, *151*, 106–116. doi:10.1016/j.actpsy.2014.06.005 PMID:24973629

Da Cruz Alves, N., Gresse Von Wangenheim, C., & Hauck, J. C. (2019). Approaches to assess computational thinking competences based on code analysis in K-12 education: A systematic mapping study. *Informatics in Education*, *18*(1), 17–39. doi:10.15388/infedu.2019.02

De Bock, D., Deprez, J., Van Dooren, W., Roelens, M., & Verschaffel, L. (2011). Abstract or concrete examples in learning mathematics? A replication and elaboration of Kaminski, Sloutsky, and Heckler's study. *Journal for Research in Mathematics Education*, *42*(2), 109–126. doi:10.5951/jresematheduc.42.2.0109

De Jong, T. (2017). Instruction based on computer simulations and virtual laboratories. In R. E. Mayer & P. A. Alexander (Eds.), *Handbook of research on learning and instruction* (2nd ed., pp. 502–521). Routledge.

Dean, A., & Lima, A. (2017). Student Experience of E-Learning Tools in HE: An Integrated Learning Framework. *European Journal of Social Sciences Education and Research*, *11*(2), 39–51. doi:10.26417/ejser.v11i2.p39-51

Dehghani, M., Acikgoz, F., Mashatan, A., & Lee, S. H. (2021). A holistic analysis towards understanding consumer perceptions of virtual reality devices in the post-adoption phase. *Behaviour & Information Technology*, 1–19.

Dekker, S., & Van Baren-Nawrocka, J. (red.) (2017). Wetenschappelijke doorbraken de klas in! Molecuulbotsingen, Stress en Taal der Zintuigen. Nijmegen: Wetenschapsknooppunt Radboud Universiteit.

Deloitte. (2017). *Soft Skills for Business Success*. Deloitte. https://www2.deloitte.com/au/en/pages/economics/articles/soft-skills-business-success.html

DeMaria, S., & Levine, A. I. (2013). The use of stress to enrich the simulated environment. In *The comprehensive textbook of healthcare simulation* (pp. 65–72). Springer. doi:10.1007/978-1-4614-5993-4_5

Dennick, R. (2016). Constructivism: Reflections on twenty-five years teaching the constructivist approach in medical education. *International Journal of Medical Education*, *7*, 200–205. doi:10.5116/ijme.5763.de11 PMID:27344115

Dewey, J. (1902). The Elementary School Teacher. *The school as social centre, 3*(2) (73-86). https://www.journals.uchicago.edu/doi/pdf/10.1086/453152

Dewey, J. (1910). Science as Subject-Matter and as Method. *Science*, *31*(787), 121–127. https://www.jstor.org/stable/1634781. doi:10.1126cience.31.787.121 PMID:17744061

Donnelly, P. J., Blanchard, N., Samei, B., Olney, A., Sun, X., Ward, B., Kelly, S., Nystrand, M., & D'Mello, S. K. (2016). Automatic Teacher Modeling from Live Classroom Audio. *UMAP '16: Proceedings of the 2016 Conference on User Modeling Adaptation and Personalization*. ACM. 10.1145/2930238.2930250

Dowell, N. M. M., Nixon, T. M., & Graesser, A. C. (2019). Group communication analysis: A computational linguistics approach for detecting sociocognitive roles in multiparty interactions. *Behavior Research Methods*, *51*(3), 1007–1041. doi:10.375813428-018-1102-z PMID:30120683

Duran, L. B., & Duran, E. (2004). The 5E Instructional Model: A Learning Cycle Approach for Inquiry-Based Science Teaching. *The Science Education Review*, *3*(2), 49–58.

Durkin, K., & Rittle-Johnson, B. (2012). *The effectiveness of using incorrect examples to support learning about decimal magnitude*. Learning and Instruction. doi:10.1016/j.learninstruc.2011.11.001

Durning, S. J., & Artino, A. R. (2011). Situativity theory: a perspective on how participants and the environment can interact: AMEE Guide no. 52. *Medical Teacher*, *33*(3), 188–199. doi:10.3109/0142159X.2011.550965 PMID:21345059

Earl, L. M. (2003). Assessment as Learning: Using Classroom Assessment to Maximize Student Learning. Thousand Oaks (CA): Corwin Press European Commission / EACEA / Eurydice, 2022. Increasing achievement and motivation in mathematics and science learning in schools. Eurydice report. Luxembourg: Publications Office of the European Union.

Efstratia, D. (2014). Experiential education through project based learning. *Procedia: Social and Behavioral Sciences*, *152*, 1256–1260. doi:10.1016/j.sbspro.2014.09.362

Ergönenç, J., Neumann, K., & Fischer, H. E. (2014). The impact of pedagogical content knowledge on cognitive activation and student learning. In H. E. Fischer, P. Labudde, K. Neumann, & J. Viiri (Eds.), *Quality of instruction in physics: Comparing Finland*. Waxmann Verlag.

Erickson, F. (2007). Ways of seeing video: Toward a phenomenology of viewing minimally edited footage. In R. Goldman, R. Pea, B. Barron, & S. J. Derry (Eds.), *Video Research in the Learning Sciences* (pp. 145–158). Lawrence Erlbaum.

Eshach, H. (2010). An analysis of conceptual flow patterns and structures in the physics classroom. *International Journal of Science Education, 32*(4), 451–477. doi:10.1080/09500690802635247

Esposito, A. G., & Bauer, P. J. (2017). Going beyond the lesson: Self-generating new factual knowledge in the classroom. *Journal of Experimental Child Psychology, 153*, 110–125. doi:10.1016/j.jecp.2016.09.003 PMID:27728784

Fadde, P., & Rich, P. (2010). Guerrilla video: A new protocol for producing classroom video. *Educational Technology, 50*(1), 4–8.

Falloon, G. (2015). What is the difference? Learning collaboratively using iPads in conventional classrooms. *Computers & Education, 84*, 62–77.

Faour, M. A., & Ayoubi, Z. (2018). The effect of using virtual laboratory on grade 10 students' conceptual understanding and their attitudes towards physics. [JESEH]. *Journal of Education in Science, Environment and Health, 4*(1), 54–68. doi:10.21891/jeseh.387482

Finnish National Board of Education. (2016). *National Core Curriculum for Basic Education 2014*. Helsinki, Finland: Finnish National Board of Education. .

Fischer, H. E., Neumann, K., Labudde, P., & Viiri, J. (Eds.). (2014). *Quality of instruction in physics: Comparing Finland*. Waxmann Verlag.

Fraillon, J., Ainley, J., Schulz, W., Friedman, T., & Duckworth, D. (2019). *Preparing for life in a digital world: IEA International Computer and Information Literacy Study 2018 International Report*. International Association for the Evaluation of Educational Achievement. doi:10.1007/978-3-030-38781-5

Franco, R. (2018). *L'IBSE: un approccio didattico vincente per l'insegnamento nelle scuole superiori. Il caso innovativo delle scienze della terra «investigate» attraverso l'arte*. Lampi di Stampa.

Fuchs, S., Gilbert, J. K., Powell, S. R., Cirino, P. T., Fuchs, D., Hamlett, C. L., Seethaler, P. M., & Tolar, T. D. (2016). The Role of Cognitive Processes, Foundational Math Skill, and Calculation Accuracy and Fluency in Word-Problem Solving versus Pre-Algebraic Knowledge. *Developmental Psychology, 52*(12), 2085–2098. doi:10.1037/dev0000227 PMID:27786534

Fulmer, G. W., Tanas, J., & Weiss, K. A. (2018). The challenges of alignment for the Next Generation Science Standards. *Journal of Research in Science Teaching, 55*(7), 1076–1100. doi:10.1002/tea.21481

Furtak, E. M., & Shavelson, R. J. (2009). Guidance, Conceptual Understanding, and Student Learning: An Investigation of Inquiry-Based Teaching in the US. In T. Janik & T. Seidel (Eds.), *The Power of Video Studies in Investigating Teaching and Learning in the Classroom* (pp. 181–206). Waxmann.

Fyfe, E. R., DeCaro, M. S., & Rittle-Johnson, B. (2015). When feedback is cognitive-demanding: The importance of working memory capacity. *Instructional Science*, *43*(1), 73–91. doi:10.100711251-014-9323-8

Gagnon, K., Young, B., Bachman, T., Longbottom, T., Severin, R., & Walker, M. J. (2020). Doctor of physical therapy education in a hybrid learning environment: Reimagining the possibilities and navigating a "new normal". *Physical Therapy*, *100*(8), 1268–1277. doi:10.1093/ptj/pzaa096 PMID:32424417

Gandolfi, E., Kosko, K. W., & Ferdig, R. E. (2021). Situating presence within extended reality for teacher training: Validation of the extended Reality Presence Scale (XRPS) in preservice teacher use of immersive 360 video. *British Journal of Educational Technology*, *52*(2), 824–841. doi:10.1111/bjet.13058

Gaudin, C., & Chaliès, S. (2015). Video viewing in teacher education and professional development: A literature review. *Educational Research Review*, *16*, 41–67. doi:10.1016/j.edurev.2015.06.001

Geier, R., Blumenfeld, P. C., Marx, R. W., Krajcik, J. S., Fishman, B., Soloway, E., & Clay-Chambers, J. (2008). Standardized test outcomes for students engaged in inquiry-based science curricula in the context of urban reform. *Journal of Research in Science Teaching*, *45*(8), 922–939. doi:10.1002/tea.20248

Geller, C., Neumann, K., Boone, W. J., & Fischer, H. E. (2014). What makes the Finnish different in science? Assessing and comparing students' science learning in three countries. *International Journal of Science Education*, *36*(18), 3042–3066. doi:10.1080/09500693.2014.950185

Gerup, J., Soerensen, C. B., & Dieckmann, P. (2020). Augmented reality and mixed reality for healthcare education beyond surgery: An integrative review. *International Journal of Medical Education*, *11*, 1–18. doi:10.5116/ijme.5e01.eb1a PMID:31955150

Gibson, J. J. (2014). The ecological approach to visual perception: classic edition. Psychology Press.

Gillies, R. (2016). Dialogic interactions in the cooperative classrooms. *International Journal of Educational Research*, *76*, 178–189. doi:10.1016/j.ijer.2015.02.009

Goh, P. S., & Sandars, J. (2020). A vision of the use of technology in medical education after the COVID-19 pandemic. *MedEdPublish*, 9.

Gong, D., Yang, H. H., & Cai, J. (2020). Exploring the key influencing factors on college students' computational thinking skills through flipped-classroom instruction. *International Journal of Educational Technology in Higher Education*, *17*(19), 19. doi:10.118641239-020-00196-0

González-Calero, J. A., Berciano, A., & Arnau, D. (2019). The role of language on the reversal error. A study with bilingual Basque-Spanish students. *Mathematical Thinking and Learning*, 1–19.

Gotwals, A. W., & Birmingham, D. (2016). Eliciting, identifying, interpreting, and responding to students' ideas: Teacher candidates' growth in formative assessment practices. *Research in Science Education, 46*(3), 365–388. doi:10.100711165-015-9461-2

Graesser, A. C., Person, N. K., & Huber, J. D. (1992). Mechanisms that generate questions. In T. E. Lauer, E. Peacock, & A. C. Graesser (Eds.), *Questions and information systems* (pp. 67–187). Erlbaum.

Graf, A. C., Jacob, E., Twigg, D., & Nattabi, B. (2020). Contemporary nursing graduates' transition to practice: A critical review of transition models. *Journal of Clinical Nursing, 29*(15-16), 3097–3107. doi:10.1111/jocn.15234 PMID:32129522

Gravemeijer, K., Stephan, M., Julie, C., Lin, F., & Ohtani, M. (2017). What mathematics education may prepare students for the society of the future? *International Journal of Science and Mathematics Education, 15*(S1), 105–123. doi:10.100710763-017-9814-6

Grossnickle, E. M., Dumas, D., Alexander, P. A., & Baggetta, P. (2016). Individual differences in the process of relational reasoning. *Learning and Instruction, 42*, 141–159. doi:10.1016/j.learninstruc.2016.01.013

Grover, S. (2018). The 5th 'C' of 21st-century skills? Try computational thinking (not coding). *EdSurge News*. https://edtechbooks.org/-Pz

Grover, S., & Pea, R. (2013). Computational thinking in K–12: A review of the state of the field. *Educational Researcher, 42*(1), 38–43. doi:10.3102/0013189X12463051

Grover, S., & Pea, R. (2018). Computational thinking: a competency whose time has come. Computer Science Education: Perspectives on teaching and learning in school (pp. 19–37). London. *Bloomsbury Academic., 1033142*. doi:10.1080/08993408.2015

Guggemos, J. (2021). On the predictors of computational thinking and its growth at the high-school level. *Computers & Education, 161*, 104060. doi:10.1016/j.compedu.2020.104060

Guyon, I., & Elisseeff, A. (2003). An introduction to variable and feature selection. *Journal of Machine Learning Research, 3*, 1157–1182.

Hafen, C. A., Hamre, B. K., Allen, J. P., Bell, C. A., Gitomer, D. H., & Pianta, R. C. (2015). Teaching through interactions in secondary school classrooms: Revisiting the factor structure and practical application of the classroom assessment scoring system–Secondary. *The Journal of Early Adolescence, 35*(5–6), 651–680. doi:10.1177/0272431614537117 PMID:28232770

Hamilton, D., McKechnie, J., Edgerton, E., & Wilson, C. (2021). Immersive virtual reality as a pedagogical tool in education: A systematic literature review of quantitative learning outcomes and experimental design. *Journal of Computers in Education, 8*(1), 1–32. doi:10.100740692-020-00169-2

Harley, W., & Elstgeest, J. (1985). *Primary Science: Taking the Plunge. How to Teach Primary Science More Effectively*. Heinemann.

Harris, C. J., Penuel, W. R., D'Angelo, C. M., DeBarger, A. H., Gallagher, L. P., Kennedy, C. A., Cheng, B. H., & Krajcik, J. S. (2015). Impact of project-based curriculum materials on student learning in science: Results of a randomized controlled trial. *Journal of Research in Science Teaching*, *52*(10), 1362–1385. doi:10.1002/tea.21263

Hartanto, A., Yang, H., & Yang, S. (2018). Bilingualism positively predicts mathematical competence: Evidence from two large-scale studies. *Learning and Individual Differences*, *61*, 216–227. doi:10.1016/j.lindif.2017.12.007

Hartman, E., Reynolds, N. P., Ferrarini, C., Messmore, N., Evans, S., Al-Ebrahim, B., & Brown, J. M. (2020). Coloniality-decoloniality and critical global citizenship: Identity, belonging, and education abroad. *Frontiers: The Interdisciplinary Journal of Study Abroad*, *32*(1), 33–59. doi:10.36366/frontiers.v32i1.433

Hawking, S. (2009). *A brief history of time: from big bang to black holes*. Random House.

Hawkins, F. H. (2017). *Human factors in flight*. Routledge. doi:10.4324/9781351218580

Hayes, C., & Capper, S. (2020). Illustrating the transcendence of disciplinarity. In *Beyond Disciplinarity* (pp. 40–49). Routledge.

Hayes, C., & Graham, Y. (2020). *Designing a Benchmarking Tool for Testing Posttest Confidence Levels in Emergency Obstetrics Training*. SAGE Publications. doi:10.4135/9781529709285

Hayes, C., Hinshaw, K., & Petrie, K. (2019). Reconceptualising medical curriculum design in strategic clinical leadership training for the 21st century physician. In *Preparing Physicians to Lead in the 21st century* (pp. 147–163). IGI Global.

Helle, L., Tynjälä, P., & Olkinuora, E. (2006). Project-based learning in post-secondary education–theory, practice and rubber sling shots. *Higher Education*, *51*(2), 287–314. doi:10.100710734-004-6386-5

Hennessy, S., Howe, C., Mercer, N., & Vrikki, M. (2020). Coding classroom dialogue: Methodological considerations for researchers. *Learning, Culture and Social Interaction*, *25*, 100404. doi:10.1016/j.lcsi.2020.100404

Hennico, J., Reuter, R. A. P., & Weinberger, M. (2021). Factors Affecting the Teaching of Computational Thinking in Fundamental Schools: A Path Analysis. A paper presented at the *15th Annual Conference of the European Association for Practitioner Research on Improving Learning*. University of Luxenberg. https://hdl.handle.net/10993/48689

Heong, Y. M., Ping, K. H., Hamdan, N., Ching, K. B., Yunos, J. M., Mohamad, M. M., ... Azid, N. (2020). Integration of Learning Styles and Higher Order Thinking Skills among Technical Students. *Journal of Technical Education and Training*, *12*(3), 171–179.

He, P., Chen, I.-C., Touitou, I., Bartz, K., Schneider, B., & Krajcik, J. (2023). Predicting student science achievement using post-unit assessment performances in a coherent high school chemistry project-based learning system. *Journal of Research in Science Teaching*, *60*(4), 724–760. doi:10.1002/tea.21815

Hiebert, J., Morris, A. K., Berk, D., & Jansen, A. (2007). Preparing teachers to learn from teaching. *Journal of Teacher Education*, *58*(1), 47–61. doi:10.1177/0022487106295726

Hilburg, R., Patel, N., Ambruso, S., Biewald, M. A., & Farouk, S. S. (2020). Medical education during the coronavirus disease-2019 pandemic: Learning from a distance. *Advances in Chronic Kidney Disease*, *27*(5), 412–417. doi:10.1053/j.ackd.2020.05.017 PMID:33308507

Hilty, D. M., Parish, M. B., Chan, S., Torous, J., Xiong, G., & Yellowlees, P. M. (2020). A comparison of in-person, synchronous and asynchronous telepsychiatry: Skills/competencies, teamwork, and administrative workflow. *Journal of Technology in Behavioral Science*, *5*(3), 273–288. doi:10.100741347-020-00137-8

Hohlfeld, T. N., Ritzhaupt, A. D., Dawson, K., & Wilson, M. L. (2017). An examination of seven years of technology integration in Florida schools: Through the lens of the levels of digital divide in Schools. *Computers & Education*, *113*, 135–161. doi:10.1016/j.compedu.2017.05.017

Hollingsworth, H., & Clarke, D. (2017). Video as a tool for focusing teacher self-reflection: Supporting and provoking teacher learning. *Journal of Mathematics Teacher Education*, *20*(5), 457–475. doi:10.100710857-017-9380-4

Horton, S. (2021). Empathy Cannot Sustain Action in Technology Accessibility. *Frontiers of Computer Science*, *3*, 31.

Hoven, J., & Garelick, B. (2007). Singapore Math: Using the bar model approach, Singapore textbooks enable students to solve complex math problems—and learn how to think symbolically. *Educational Leadership*, *65*, 38–21.

Howe, C., Hennessy, S., Mercer, N., Vrikki, M., & Wheatley, L. (2019). Teacher–student dialogue during classroom teaching: Does it really impact on student outcomes? *Journal of the Learning Sciences*, *28*(4-5), 462–512. doi:10.1080/10508406.2019.1573730

Howell, H., & Mikeska, J. N. (2021). Approximations of practice as a framework for understanding authenticity in simulations of teaching. *Journal of Research on Technology in Education*, *53*(1), 8–20. doi:10.1080/15391523.2020.1809033

Huang, G. Y., Chen, J., Liu, H., Fu, W., Ding, W., Tang, J., Yang, S., Li, G., & Liu, Z. (2020). Neural multi-task learning for teacher question detection in online classrooms. https://arxiv.org/abs/2005.07845v1 doi:10.1007/978-3-030-52237-7_22

Humpherys, S. L., Bakir, N., & Babb, J. (2021). Experiential learning to foster tacit knowledge through a role play, business simulation. *Journal of Education for Business*, 1–7.

Inkinen, J., Klager, C., Juuti, K., Schneider, B., Salmela-Aro, K., Krajcik, J., & Lavonen, J. (2020). High school students' situational engagement associated with scientific practices in designed science learning situations. *Science Education*, *104*(4), 667–692. doi:10.1002ce.21570

ISTE & CSTA. (2011). Operational definition of computational thinking for K–12 Education. ISTE. Indiana Government. http://www.iste.org/docs/ctdocuments/computational-thinking-operational-definition flyer.pdf

Jacobs, J., Scornavacco, K., Harty, C., Suresh, A., Lai, V., & Sumner, T. (2022). Promoting rich discussions in mathematics classrooms: Using personalized, automated feedback to support reflection and instructional change. *Teaching and Teacher Education*, *112*, 103631. doi:10.1016/j.tate.2022.103631

Jenner, K. (n.d.). *Indiana Academic Standards: Science and Engineering Practices.* Department of Education. https://media.doe.in.gov/news/science-and-engineering-practices.pdf

Jentsch, F., & Curtis, M. (2017). *Simulation in aviation training*. Routledge. doi:10.4324/9781315243092

Johnson, C. E., Kimble, L. P., Gunby, S. S., & Davis, A. H. (2020). Using deliberate practice and simulation for psychomotor skill competency acquisition and retention: A mixed-methods study. *Nurse Educator*, *45*(3), 150–154. doi:10.1097/NNE.0000000000000713 PMID:31246693

Jongerius, C., Hessels, R. S., Romijn, J. A., Smets, E. M., & Hillen, M. A. (2020). The measurement of eye contact in human interactions: A scoping review. *Journal of Nonverbal Behavior*, *44*(3), 1–27. doi:10.100710919-020-00333-3

Juraschek, M., Büth, L., Posselt, G., & Herrmann, C. (2018). Mixed reality in learning factories. *Procedia Manufacturing*, *23*, 153–158. doi:10.1016/j.promfg.2018.04.009

Kääntä, L. (2015). The multimodal organization of teacher-led classroom interaction. In C. Jenks & P. Seedhouse (Eds.), *International perspectives on ELT classroom interaction* (pp. 64–83). Palgrave Macmillan. doi:10.1057/9781137340733_5

Kaminski, J. A., & Sloutsky, V. M. (2013). Extraneous perceptual information interferes with children's acquisition of mathematical knowledge—. *Journal of Educational Psychology*, *105*(2), 351–363. doi:10.1037/a0031040

Kandemir, M. A., & Karadeniz, I. (2021). Preservice teachers' cognitive and metacognitive processes in integrated STEM modeling activity. [JESEH]. *Journal of Education in Science, Environment and Health*, *7*(2), 104–127. doi:10.21891/jeseh.832574

Kang, J., Diederich, M., Lindgren, R., & Junokas, M. (2021). Gesture patterns and learning in an embodied XR science simulation. *Journal of Educational Technology & Society*, *24*(2), 77–92.

Kanimozhi. (2018). Virtual Learning Environment. *International Journal of Advanced Research in Computer Science, 9*(1), 11–14.

Karunathilake, I. M., & Samarasekera, D. D. (2021). Learning In The 21st Century— 'What's All the Fuss about Change?'. In Educate, Train and Transform: Toolkit on Medical and Health Professions Education (pp. 1-14). Routledge.

Kaseva, T., Rouhe, A., & Kurimo, M. (2019). SphereDiar: An effective speaker diarization system for meeting data. Annual Conference of the International Speech Communication Association. ISCA.

Kauffman, H. (2015). A review of predictive factors of student success in and satisfaction with online learning. *Research in Learning Technology*. doi:. v23. 26507. doi:10.3402/rlt

Kay, R., & Edwards, J. (2012). A formative analysis of the use of worked example video podcasts in middle school mathematics classrooms. *Canadian Journal of Learning and Technology, 38*(3). doi:10.21432/T2PK5Z

Kelly, S., Olney, A. M., Donnelly, P., Nystrand, M., & D'Mello, S. K. (2018). Automatically measuring question authenticity in real-world classrooms. *Educational Researcher, 47*(7), 451–464. doi:10.3102/0013189X18785613

King, O., Borthwick, A., Nancarrow, S., & Grace, S. (2018). Sociology of the professions: What it means for podiatry. *Journal of Foot and Ankle Research, 11*(1), 1–8. doi:10.118613047-018-0275-0 PMID:29942353

Kizilcec, R. F., Piech, C., & Schneider, E. (2017). Deconstructing disengagement: Analyzing learner subpopulations in massive open online courses. In Proceedings of the Fourth (2017) *ACM Conference on Learning @ Scale* (pp. 303-312). ACM.

Kloser, M. (2014). Identifying a core set of science teaching practices: A Delphi expert panel approach. *Journal of Research in Science Teaching, 51*(9), 1185–1217. doi:10.1002/tea.21171

Knopke, V. (2019). Where are all the students? Factors that encourage female participants in technology education. In P. Williams & D. Barlex (Eds.), *Explorations in Technology Education Research. Contemporary Issues in Technology Education* (pp. 223–239). Springer. doi:10.1007/978-981-13-3010-0_15

Kotsopoulos, D., Floyd, L., Khan, S., Namukasa, I. K., Somanath, S., Weber, J., & Yiu, C. (2017). A pedagogical framework for computational thinking. *Digital Experiences in Mathematics Education, 3*(2), 154–171. doi:10.100740751-017-0031-2

Koufidis, C., Manninen, K., Nieminen, J., Wohlin, M., & Silén, C. (2021). Unravelling the polyphony in clinical reasoning research in medical education. *Journal of Evaluation in Clinical Practice, 27*(2), 438–450. doi:10.1111/jep.13432 PMID:32573080

Koul, R., Sheffield, R., & McIlvenny, L. (2021). *Teaching 21st Century Skills*. Springer. https://link.springer.com/book/10.1007/978-981-16-4361-3

Koul, R. B., Fraser, B. J., Maynard, N., & Tade, M. (2017). Evaluation of engineering and technology activities in primary schools in terms of learning environment, attitudes and understanding. *Learning Environments Research*. doi:10.100710984-017-9255-8

Krajcik, J. S., & Shin, N. (2014). Project-based learning. In R. K. Sawyer (Ed.), The Cambridge handbook of learning sciences, (2nd ed.), New York: Cambridge. doi:10.1017/CBO9781139519526.018

Krajcik, J. S., Blumenfeld, P. C., Marx, R. W., & Soloway, E. (1994). A collaborative model for helping middle grade science teachers learn project-based instruction. *The Elementary School Journal, 94*(5), 483–497. doi:10.1086/461779

Krajcik, J. S., & Czerniak, C. M. (2018). *Teaching science in elementary and middle school: A project-based learning approach*. Routledge. doi:10.4324/9781315205014

Krajcik, J. S., Palincsar, A., & Miller, E. (2015). *Multiple literacy in project-based learning*. Michigan State University.

Krajcik, J. S., Schneider, B., Miller, E., Chen, I. C., Bradford, L., Peek-Brown, D., Bartz, K., Baker, Q., Palicsar, A., Miller, C., & Codere, S. (2023). Assessing the Effect of Project-Based Learning on Science Learning in Elementary Schools. *American Educational Research Journal*, *60*(1), 70–102. doi:10.3102/00028312221129247

Krajcik, J. S., & Shin, N. (2023). *Student Conceptions*. Conceptual Change, and Learning Progressions.

Krajcik, J., McNeill, K. L., & Reiser, B. J. (2008). Learning-goals-driven design model: Developing curriculum materials that align with national standards and incorporate project-based pedagogy. *Science Education*, *92*(1), 1–32. doi:10.1002ce.20240

Kulgemeyer, C., & Schecker, H. (2014). Research on Educational Standards in German Science Education - Towards a model of student competences. *Eurasia Journal of Mathematics, Science and Technology Education*, *10*(4), 257–269. doi:10.12973/eurasia.2014.1081a

Kupers, E., van Dijk, M., & van Geert, P. (2015). Within-teacher differences in one-to-one teacher-student interactions in instrumental music lessons. *Learning and Individual Differences*, *37*, 283–289. doi:10.1016/j.lindif.2014.11.012

Lam, D. S. H., & Chan, K. K. H. (2020). Characterising pre-service secondary science teachers' noticing of different forms of evidence of student thinking. *International Journal of Science Education*, *42*(4), 576–597. doi:10.1080/09500693.2020.1717672

Lampert, M., Franke, M. L., Kazemi, E., Ghousseini, H., Turrou, A. C., Beasley, H., & Crowe, K. (2013). Keeping it complex: Using rehearsals to support novice teacher learning of ambitious teaching. *64*(3), 226-243. https://journals.sagepub.com/doi/abs/10.1177/0022487112473837

Lämsä, J., Uribe, P., Jiménez, A., Caballero, D., Hämäläinen, R., & Araya, R. (2021). Deep Networks for Collaboration Analytics: Promoting Automatic Analysis of Face-to-Face Interaction in the Context of Inquiry-Based Learning. *Journal of Learning Analytics*, *8*(1), 113–125. doi:10.18608/jla.2021.7118

Larmer, J., Mergendoller, J., & Boss, S. (2015). *Setting the standard for project-based learning*. ASCD.

Lee, C., & Chen, M. (2015). Effects of worked examples using manipulatives on fifth graders' learning performance and attitude toward mathematics. *Journal of Educational Technology & Society*, *18*(1), 264-275.40

Lee, V. S., Greene, D. B., Odom, J., Schechter, E., & Slatta, R. W. (2004). What is inquiryguided learning. In V. S. Lee (Ed.), Teaching and learning through inquiry: A guidebook for institutions and instructors (pp. 3-15). Sterling, VA: Stylus Publishing.

Lee, I., Martin, F., Denner, J., Coulter, B., Allan, W., Erickson, J., Malyn-Smith, J., & Werner, L. (2011). Computational thinking for youth in practice. *ACM Inroads*, *2*(1), 32–37. doi:10.1145/1929887.1929902

Lee, N. H., Lee, J., & Wong, Z. Y. (2021). Preparing Students for the Fourth Industrial Revolution through Mathematical Learning: The Constructivist Learning Design. *Korean J. Opt. Photon*, *31*(3), 321–356. doi:10.29275/jerm.2021.31.3.321

Lee, O., & Grapin, S. E. (2022). The role of phenomena and problems in science and STEM education: Traditional, contemporary, and future approaches. *Journal of Research in Science Teaching*, *59*(7), 1301–1309. doi:10.1002/tea.21776

Lehesvuori, S., & Ametller, J. (2021). Exploring coherence and authorship in pedagogical link-making in science. *International Journal of Science Education*, *43*(17), 2791–2813. doi:10.1080/09500693.2021.1991599

Lehesvuori, S., Ketonen, L., & Hähkiöniemi, M. (2022). Utilizing Informal Formative Assessment and Dialogicity During Reflections on Educational Dialogue in Mathematics. *Studia Paedagogica*, *27*(2), 55–75. doi:10.5817/SP2022-2-3

Lehesvuori, S., Viiri, J., & Rasku-Puttonen, H. (2011). Introducing Dialogic Teaching to Science Student Teachers. *Journal of Science Teacher Education*, *22*(8), 705–727. doi:10.100710972-011-9253-0

Lehesvuori, S., Viiri, J., Rasku-Puttonen, H., Moate, J., & Helaakoski, J. (2013). Visualizing communication structures in science classrooms: Tracing cumulativity in teacher-led whole class discussions. *Journal of Research in Science Teaching*, *50*(8), 912–939. doi:10.1002/tea.21100

Lemke, J. L. (1990). *Talking science: Language, learning and values*. Ablex Publishing Company.

Levin, D. M., & Richards, J. (2011). Learning to attend to the substance of students' thinking in science. *Science Educator*, *20*(2), 1.

Levitt, H. M. (2021). Qualitative generalisation, not to the population but to the phenomenon: Reconceptualising variation in qualitative research. *Qualitative Psychology*, *8*(1), 95–110. doi:10.1037/qup0000184

Lewis, K. (2020). Technology in the workplace: Redefining skills for the 21st century. *The Midwest Quarterly*, *61*(3), 348–355.

Lewis, T. (1998, Fall). Petrina, s. and Hill, A. M. (1998) Problem Posing-Adding a Creative Increment to Technological Problem solving. *Journal of Industrial Teacher Education*, *36*(1). https://scholar.lib.vt.edu/ejournals/JITE/v36n1/lewis.html

Li, T., Miller, E., Chen, I.C., Bartz, K., Codere, S., & Krajcik, J. (2021). The relationship between teacher's support of literacy development and elementary students' modelling proficiency in project-based learning science classrooms. *Education 3-13: International Journal of Primary, Elementary and Early Years Education.* . [ESCI]. doi:10.1080/03004279.2020.1854959

Limniou, M., & Smith, M. (2010). Teachers' and students' perspectives on teaching and learning through virtual learning environments. *European Journal of Engineering Education, 35*(6), 645–653. doi:10.1080/03043797.2010.505279

Li, T., Chen, I., Miller, E., Miller, C., Schneider, B., & Krajcik, J. (Accepted). The Relationships between Elementary Students' Knowledge-In-Use Performance and Their Science Achievement. *Journal of Research in Science Teaching.*

Littleton, K., & Mercer, N. (2013). *Interthinking: Putting talk to work.* Routledge. doi:10.4324/9780203809433

Loeng, S. (2018). Various ways of understanding the concept of andragogy. *Cogent Education, 5*(1), 1496643. doi:10.1080/2331186X.2018.1496643

Logeswaran, A., Munsch, C., Chong, Y. J., Ralph, N., & McCrossnan, J. (2021). The role of extended reality technology in healthcare education: Towards a learner-centred approach. *Future Healthcare Journal, 8*(1), e79–e84. doi:10.7861/fhj.2020-0112 PMID:33791482

Luctkar-Flude, M., & Tyerman, J. (2021). The Rise of Virtual Simulation: Pandemic Response or Enduring Pedagogy? *Clinical Simulation in Nursing, 57*, 1–2. doi:10.1016/j.ecns.2021.06.008

Luna, M. J., & Sherin, M. G. (2017). Using a video club design to promote teacher attention to students' ideas in science. *Teaching and Teacher Education, 66*, 282–294. doi:10.1016/j.tate.2017.04.019

Luo, C., Lan, Y., Luo, X. R., & Li, H. (2021). The effect of commitment on knowledge sharing: An empirical study of virtual communities. *Technological Forecasting and Social Change, 163*, 120438. doi:10.1016/j.techfore.2020.120438

Lv, K., Sun, Z., & Xu, M. (2021). Artificial Intelligent Based Video Analysis on the Teaching Interaction Patterns in Classroom Environment. *International Journal of Information and Education Technology (IJIET), 11*(3), 126–130. doi:10.18178/ijiet.2021.11.3.1500

Makrides, G. A. (2019). The evolution of education from Education 1.0 to Education 4.0: Is it an evolution or a revolution? *Cloud.* https://www.l-cloud.eu/wp-content/uploads/2019/03/Evolution_of_Education.pdf

Maraver, M. J., Bajo, M. T., & Gomez-Ariza, C. J. (2016). Training on Working Memory and Inhibitory Control in Young Adults. *Frontiers in Human Neuroscience, 10*, 1–18. doi:10.3389/fnhum.2016.00588 PMID:27917117

Marconato, G., Molinari, M., & Musco, F. (2016). *Competenti si diventa.* Pearson Longman.

Mathew, P. S., & Pillai, A. S. (2020). Role of Immersive (XR) Technologies in Improving Healthcare Competencies: A Review. *Virtual and Augmented Reality in Education, Art, and Museums*, 23-46.

Mayer, R. E. (2011). Instruction based on visualisations. In *Handbook of research on learning and instruction* (pp. 441–459). Routledge. doi:10.4324/9780203839089-30

Mcgowan, V. C., Klein, E., & Morrison, D. (2019). *Teaching Tools for Science, Technology, Engineering and Math (STEM) Education.* STEM Teaching Tools. https://stemteachingtools.org/brief/56

McGrath, J. L., Taekman, J. M., Dev, P., Danforth, D. R., Mohan, D., Kman, N., ... Won, K. (2018). Using virtual reality simulation environments to assess competence for emergency medicine learners. *Academic Emergency Medicine*, *25*(2), 186–195. doi:10.1111/acem.13308 PMID:28888070

McKinsey Global Institute. (2017). *Skill Shift: Automation and the Future of the Workforce*. McKinsey. https://www.mckinsey.com/featured-insights/future-of-work/skill-shift-automation-and-the-future-of-the-workforce

McLaren, B. M., Adams, D. M., & Mayer, R. E. (2015). Delayed learning effects with erroneous examples: A study of learning decimals with a web-based tutor. *International Journal of Artificial Intelligence in Education*, *25*(4), 520–542. doi:10.100740593-015-0064-x

Melnyk, B. M., Tan, A., Hsieh, A. P., Gawlik, K., Arslanian-Engoren, C., Braun, L. T., Dunbar, S., Dunbar-Jacob, J., Lewis, L. M., Millan, A., Orsolini, L., Robbins, L. B., Russell, C. L., Tucker, S., & Wilbur, J. (2021). Critical care nurses' physical and mental health, worksite wellness support, and medical errors. *American Journal of Critical Care*, *30*(3), 176–184. doi:10.4037/ajcc2021301 PMID:34161980

Mercer, N., Dawes, L., & Staarman, J. K. (2009). Dialogic teaching in the primary science classroom. *Language and Education*, *23*(4), 353–369. doi:10.1080/09500780902954273

Merriam, S., & Tisdell, E. (2016). Qualitative research: A guide to design and implementation (4th ed.) John Wiley & Sons, Inc.

Merritt, J., Lee, M. Y., Rillero, P., & Kinach, B. M. (2017). Problem-based learning in K-8 mathematics and science education: A literature review. *The Interdisciplinary Journal of Problem-Based Learning*, *11*(2), 3. doi:10.7771/1541-5015.1674

Mielicki, M. K., Kacinik, N. A., & Wiley, J. (2017). Bilingualism and symbolic abstraction: Implications for algebra learning. *Learning and Instruction*, *49*, 242–250. doi:10.1016/j.learninstruc.2017.03.002

Miller, E., & Reigh, E. (2020). *Interactional approach to language: Supporting all students to engage in equitable sensemaking.*

Miller, E. C., & Krajcik, J. S. (2019). Promoting deep learning through project-based learning: A design problem. *Disciplinary and Interdisciplinary Science Education Research*, *1*(1), 1–10. doi:10.118643031-019-0009-6

Miller, E. C., Severance, S., & Krajcik, J. (2020). Connecting Computational Thinking and Science in a US Elementary Classroom. In J. Anderson & Y. Li (Eds.), *Integrated Approaches to STEM Education: An International Perspective* (pp. 185–204). Springer, Cham. doi:10.1007/978-3-030-52229-2_11

Ministero dell'Istruzione, dell'Università e della Ricerca (MIUR). (2012). Indicazioni nazionali per il curricolo della scuola dell'infanzia [num. spec.]. *Annali della pubblica istruzione, LXXXVIII.*

Ministry of Education, (People's Republic of China). (2003). National standards of general senior secondary school chemistry curriculum. *People's Education Press.* (in Chinese)

Ministry of Science and Technology and Publicity Department. (2016). *The Chinese Citizens Science Quality Benchmark.* Ministry of Science and Technology and the Publicity Department. http://www. gov.cn/gongbao/content/2016/content_5103155.htm

Mitchell, R., & Boyle, B. (2021). Understanding the role of profession in multidisciplinary team innovation: Professional identity, minority dissent and team innovation. *British Journal of Management, 32*(2), 512–528. doi:10.1111/1467-8551.12419

Moreira, D. (2020). Virtual networks and asynchronous communities: methodological reflections on the digital. In *Ethnography in Higher Education* (pp. 177–196). Springer VS. doi:10.1007/978-3-658-30381-5_11

Mortimer, E. F., & Machado, A. H. (2000). Anomalies and conflicts in classroom discourse. *Science Education, 84*(4), 429–444. doi:10.1002/1098-237X(200007)84:4<429::AID-SCE1>3.0.CO;2-#

Mortimer, E. F., & Scott, P. (2003). *Meaning making in science classrooms.* Open University Press.

Mortimore, G., Reynolds, J., Forman, D., Brannigan, C., & Mitchell, K. (2021). From expert to advanced clinical practitioner and beyond. *British Journal of Nursing (Mark Allen Publishing), 30*(11), 656–659. doi:10.12968/bjon.2021.30.11.656 PMID:34109817

Muhonen, H., Pakarinen, E., Poikkeus, A.-M., Lerkkanen, M.-K., & Rasku-Puttonen, H. (2018). Quality of educational dialogue and association with students' academic performance. *Learning and Instruction, 55*, 67–79. doi:10.1016/j.learninstruc.2017.09.007

Müller, K., Prenzel, M., Seidel, T., Schiepe-Tiska, A., & Kjærnsli, M. (2016). Science teaching and learning in schools: Theoretical and empirical foundations for investigating classroom-level processes. In S. Kuger, E. Klieme, N. Jude, & D. Kaplan (Eds.), *Assessing contexts of learning, methodology of educational measurement and assessment* (pp. 423–446). Springer. doi:10.1007/978-3-319-45357-6_17

Mullis, I. V., Martin, M. O., Foy, P., & Arora, A. (2012). TIMSS 2011 international results in mathematics. International Association for the Evaluation of Educational Achievement.

National Academies of Sciences, Engineering, and Medicine. (2019). *Science and Engineering for Grades 6-12: Investigation and Design at the Center.* Washington, DC: The National Academies Press.

National Research Council. (2010). *Report of a Workshop on the Scope and Nature of Computational Thinking.* Washington, DC: The National Academies Press. https://doi.org/10.17226/12840

National Research Council. (2011). *Assessing 21st century skills: Summary of a workshop.* The National Academies Press.

National Research Council. (2012a). *A Framework for K-12 science education: practices, crosscutting concepts, and core ideas*. The National Academies Press.

National Research Council. (2012b). *Education for life and work: Developing transferable knowledge and skills in the 21st century*. National Academies Press.

Negev, M., Dahdal, Y., Khreis, H., Hochman, A., Shaheen, M., Jaghbir, M. T., Alpert, P., Levine, H., & Davidovitch, N. (2021). Regional lessons from the COVID-19 outbreak in the Middle East: From infectious diseases to climate change adaptation. *The Science of the Total Environment*, *768*, 144434. doi:10.1016/j.scitotenv.2020.144434 PMID:33444865

Newman, S. D. (2016). *Does finger sense predict addition performance?* Cognitive Psychology. doi:10.100710339-016-0756-7

Next Generation Science Standards Lead States. (2013). *Next generation science standards: For states, by states.* Next Generation Science Standards.

Nikula, T. (2005). English as an object and tool of study in classrooms: Interactional effects and pragmatic implications. *Linguistics and Education*, *16*(1), 27–58. doi:10.1016/j.linged.2005.10.001

Niss, M., & Højgaard, T. (2019). Mathematical competencies revisited. *Educational Studies in Mathematics*, *102*(1), 9–28. doi:10.100710649-019-09903-9

Nystrand, M., Gamoran, A., Kachur, R., & Prendergast, C. (1997). *Opening dialogue: Understanding the dynamics of language and learning in the English classroom.* Teachers College Press.

O'Connor, C., & Michaels, S. (2019). Supporting teachers in taking up productive talk moves: The long road to professional learning at scale. *International Journal of Educational Research*, *97*, 166–175. doi:10.1016/j.ijer.2017.11.003

Oberauer, K. (2019). Working Memory and Attention – A Conceptual Analysis and Review. *Journal of Cognition*, *2*(1), 1–23. doi:10.5334/joc.58

Obrad, C. (2020). Constraints and consequences of online teaching. *Sustainability (Basel)*, *12*(17), 6982. doi:10.3390u12176982

Ocloo, J., Garfield, S., Franklin, B. D., & Dawson, S. (2021). Exploring the theory, barriers and enablers for patient and public involvement across health, social care and patient safety: A systematic review of reviews. *Health Research Policy and Systems*, *19*(1), 1–21. doi:10.118612961-020-00644-3 PMID:33472647

OECD. (2016). PISA 2015. In *Assessment and analytical framework: Science, reading, mathematic and financial literacy*. OECD Publishing.

Ogegbo, A. A., & Ramnarain, U. (2022). A systematic review of computational thinking in science classrooms. *Studies in Science Education*, *58*(2), 203–230. doi:10.1080/03057267.2021.1963580

Okoye, K., Rodriguez-Tort, J. A., Escamilla, J., & Hosseini, S. (2021). Technology-mediated teaching and learning process: A conceptual study of educators' response amidst the Covid-19 pandemic. *Education and Information Technologies, 26*(6), 1–33. doi:10.100710639-021-10527-x PMID:34025205

Oliveira, A. W. (2010). Improving teacher questioning in science inquiry discussions through professional development. *Journal of Research in Science Teaching, 447*(4), 422–453. doi:10.1002/tea.20345

Olney, A. M., Kelly, S., Samei, B., Donnelly, P., & D'Mello, S. K. (2017). Assessing teacher questions in classrooms. In R. Sottilaire, A. Graesser, X. Hu, & G. Goodwin (Eds.), *Design recommendations for intelligent tutoring systems: Assessment* (Vol. 5, pp. 261–274). U.S. Army Research Laboratory.

Organisation for Economic Co-operation and Development (OECD). (2020). *Back to the Future: Four scenarios for schooling.* OECD. https://www.oecd.org/education/back-to-the-future-s-of-education-178ef527-en.htm

Organisation for Economic Cooperation and Development. (2018). *The future of education and skills: Education 2030.* OECD.

Orr, N., Matthews, B., See, Z. S., Burrell, A., Day, J., & Seengal, D. (2021). Transdisciplinarity in extended reality (XR) research design: Technological transformation and social good (co-creation session at XR+ Creativity Symposium, University of Newcastle, 2020). *Virtual Creativity, 11*(1), 163-179.

Overoye, A. L., & Storm, B. C. (2015). Harnessing the power of uncertainty to enhance learning. *Translational Issues in Psychological Science, 1*(2), 140–148. doi:10.1037/tps0000022

Owens, K. P. (2021, July). Competency-Based Experiential-Expertise and Future Adaptive Learning Systems. In *International Conference on Human-Computer Interaction* (pp. 93-109). Springer, Cham. 10.1007/978-3-030-77873-6_7

Oyewo, O. A., Ramaila, S., & Mavuru, L. (2022). Harnessing Project-Based Learning to Enhance STEM Students' Critical Thinking Skills Using Water Treatment Activity. *Education Sciences, 12*(11), 780. doi:10.3390/educsci12110780

Papert, S. (1980). *Mindstorms. Children, Computers and Powerful Ideas.* Basic books.

Park, C., & Kim, D. G. (2020). Exploring the roles of social presence and gender difference in online learning. *Decision Sciences Journal of Innovative Education, 18*(2), 291–312. doi:10.1111/dsji.12207

Parsons, S. A., Vaughn, M., Scales, R. Q., Gallagher, M. A., Parsons, A. W., Davis, S. G., Pierczynski, M., & Allen, M. (2018). Teachers' instructional adaptations: A research synthesis. *Review of Educational Research, 88*(2), 205–242. doi:10.3102/0034654317743198

Pashler, H., Bain, P., Bottge, B., Graesser, A., Koedinger, K., McDaniel, M., & Metcalfe, J. (2007). *Organising instruction and study to improve student learning: IES practice guide (NCER 2007-2004).* National Center for Education Research.

Peel, A., Sadler, T. D., & Friedrichsen, P. (2022). Algorithmic Explanations: An Unplugged Instructional Approach to Integrate Science and Computational Thinking. *Journal of Science Education and Technology, 31*(4), 428–441. doi:10.100710956-022-09965-0 PMID:35535350

Peeters, M. H. J., & Meijer, W. (2014). Onderzoekend leren: Hoe stel je een onderzoeksvraag op? *Jeugd in School en Wereld, 5*(9), 6–9.

Pellegrino, J. W., & Hilton, M. L. (2012). *Committee on defining deeper learning and 21st century skills.* Center for Education.

Penn, M., & Umesh, R. (2019). The use of virtual learning environments and achievement in physics content tests. In *Proceedings of the International Conference on Education and New Development* (pp. 493-497). ACM.

Penuel, W. R., Reiser, B. J., McGill, T. A., Novak, M., Van Horne, K., & Orwig, A. (2022). Connecting student interests and questions with science learning goals through project-based storylines. *Disciplinary and Interdisciplinary Science Education Research, 4*(1), 1–27. doi:10.118643031-021-00040-z

Peters-Burton, E., Rich, P. J., Kitsantas, A., Laclede, L., & Stehle, S. M. (2021). High School Science Teacher Use of Planning Tools to Integrate Computational Thinking. *Journal of Science Teacher Education.* doi:10.1080/1046560X.2021.1970088

Pethrick, H., Nowell, L., & Oddone Paolucci, E. (2017). Psychosocial and career outcomes of peer mentorship in medical resident education: a systematic review protocol. *Systematic Reviews, 6.* https://www.ncbi.nlm.nih.gov/pubmed/28859683

Piaget, J. (1985) *The equilibration of cognitive structures: The central problem of intellectual development.* University of Chicago Press (original work published 1975).

Piaget, J. (1936). *Origins of intelligence in the child.* Routledge & Kegan Paul.

Piaget, J. (1980). *Experiments in contradictions.* University of Chicago Press.

Pianta, R. C., Hamre, B. K., & Mintz, S. (2012). *Classroom assessment scoring system – Secondary manual.* Teachstone.

Potvin, A. S., Miller, E. A., Kuck, R., Berland, L. K., Boardman, A. G., Kavanagh, S. S., Clark, T. L., & Cheng, B. H. (2022). Mapping Enabling Conditions for High-Quality PBL: A Collaboratory Approach. *Education Sciences, 12*(3), 222. doi:10.3390/educsci12030222

Qader, W., Ameen, M., & Ahmed, B. (2019). *An overview of Bag of Words; Importance, implementation, applications, and challenges. International Engineering Conference.* 10.1109/IEC47844.2019.8950616

Rahayu, T., & Osman, K. (2019). Knowledge level and self-confidence on the computational thinking skills among science teacher candidates. *Jurna Ilm. Pendidik. Fis. Al-Biruni, 8*(1), 117–126. doi:10.24042/jipfalbiruni.v8i1.4450

Ramakrishnan, A., Ottmar, E., LoCasale-Crouch, J., & Whitehill, J. (2019). Toward automated classroom observation: Predicting positive and negative climate. *14th IEEE Int. Conf. on Automatic Face and Gesture Recognition (FG 2019)*. IEEE. 10.1109/FG.2019.8756529

Rand Corporation. (2012). Teaching and Learning 21st Century Skills: Lessons from the learning sciences. Asia Society. http://asiasociety.org/files/rand-1012report.pdf

Raver, C., & Blair, C. (2016). Neuroscientific Insights: Attention, Working Memory, and Inhibitory Control. *The Future of Children, 26*(2), 95–118. doi:10.1353/foc.2016.0014

Reiser, B. J. (2014). *Designing coherent storylines aligned with NGSS for the K-12 classroom.* Boston: National Science Education Leadership Association http://bit.ly/1TZeSaG

Reisoğlu, I., Topu, B., Yılmaz, R., Karakuş Yılmaz, T., & Göktaş, Y. (2017). 3D virtual learning environments in education: A meta-review. *Asia Pacific Education Review, 18*(1), 81–100. doi:10.100712564-016-9467-0

Richland, L. E., & McDonough, I. M. (2010). Learning by analogy: Discriminating between potential analogues. *Contemporary Educational Psychology, 35*, 28–43.

Richland, L. E., & Begolli, K. (2016). Teaching Mathematics by Comparison: Analog Visibility as a Double-Edged Sword. *Journal of Educational Psychology, 108*(2), 194–213. doi:10.1037/edu0000056

Rich, P. J., Jones, B., Belikov, O., Yoshikawa, E., & Perkins, M. (2017). Computing and engineering in elementary school: The effect of year-long training on elementary teacher self-efficacy and beliefs about teaching computing and engineering. *International Journal of Computer Science Education in Schools, 1*(1), 1–20. doi:10.21585/ijcses.v1i1.6

Rittle-Johnson, B., Star, J. R., & Durkin, K. (2012). Developing procedural flexibility: Are novices prepared to learn from comparing procedures? *The British Journal of Educational Psychology, 8*(3), 436–455. doi:10.1111/j.2044-8279.2011.02037.x PMID:22881048

Robert, I. V. (2021). Formation and development of digital transformation of domestic education on the basis of systemic convergence of pedagogical science and technology. In *SHS Web of Conferences* (*Vol. 101*, p. 03017). EDP Sciences. 10.1051hsconf/202110103017

Robinson, V. M. J. (2010). From Instructional Leadership-to-Leadership Capabilities: Empirical Findings and Methodological Challenges. *Leadership and Policy in Schools, 9*(1), 1–26. doi:10.1080/15700760903026748

Rocard, M. (2007). *Science Education Now: A Renewed Pedagogy for the Future of Europe*, Rapporto EU22-845, Bruxelles. https://ec.europa.eu/research/sciencesociety/document_library/pdf_06/report-rocard-on-science-education_en.pdf

Roller, S. A. (2016). What they notice in video: A study of prospective secondary mathematics teachers learning to teach. *Journal of Mathematics Teacher Education, 19*(5), 477–498. doi:10.100710857-015-9307-x

Román-González, M., Moreno-León, J., & Robles, G. (2019). Combining assessment tools for a comprehensive evaluation of computational thinking interventions. In S-C, Kong & H. Abelson (Eds.), Computational thinking education (pp. 79-98). Singapore: Springer. doi:10.1007/978-981-13-6528-7_6

Rothstein, D., & Santana, L. (2013). *Make just one change: teach students to ask their own questions*. Harvard Education Press.

Roussin, C. J., & Weinstock, P. (2017). SimZones: An organisational innovation for simulation programs and centers. *Academic Medicine, 92*(8), 1114–1120. doi:10.1097/ACM.0000000000001746 PMID:28562455

Ruiz-Primo, M. A. (2011). Informal formative assessment: The role of instructional dialogues in assessing students' learning. *Studies in Educational Evaluation, 37*(1), 15–24. doi:10.1016/j.stueduc.2011.04.003

Ruthotto, I., Kreth, Q., Stevens, J., Trively, C., & Melkers, J. (2020). Lurking and participation in the virtual classroom: The effects of gender, race, and age among graduate students in computer science. *Computers & Education, 151*, 103854. doi:10.1016/j.compedu.2020.103854

Saidin, N. D., Khalid, F., Martin, R., Kuppusamy, Y., & Munusamy, N. A. (2021). Benefits and challenges of applying computational thinking in education. *International Journal of Information and Education Technology (IJIET), 11*(5), 248–254. doi:10.18178/ijiet.2021.11.5.1519

Salmon, A. K., & Barrera, M. X. (2021). Intentional questioning to promote thinking and learning. *Thinking Skills and Creativity, 40*, 100822. doi:10.1016/j.tsc.2021.100822

Sanders, M. E. (2015). *The Original "Integrative STEM Education" Definition: Explained.* Vtech Works. https://vtechworks.lib.vt.edu/handle/10919/51624

Sands, P., Yadav, A., & Good, J. (2018). Computational thinking in K-12: In-service teacher perceptions of computational thinking. In M. S. Khine (Ed.), *Computational thinking in the STEM disciplines* (pp. 151–164). Springer. doi:10.1007/978-3-319-93566-9_8

Santagata, R., & Guarino, J. (2011). Using video to teach future teachers to learn from teaching. *ZDM Mathematics Education, 43*(1), 133–145. doi:10.100711858-010-0292-3

Schlotterbeck, D., Araya, R., Caballero, D., Jiménez, A., Lehesvuori, S., & Viiri, J. (2020). Assessing Teacher's Discourse Effect on Students' Learning: A Keyword Centrality Approach. In C. Alario-Hoyos, M. J. Rodríguez-Triana, M. Scheffel, I. Arnedillo-Sánchez, & S. M. Dennerlein (Eds.), EC-TEL 2020: Addressing Global Challenges and Quality Education (pp. 102-116).

Schlotterbeck, D., Jiménez, A., Araya, R., Caballero, D., Uribe, P., & Van der Molen Moris, J. (2022). "Teacher, Can You Say It Again?" Improving Automatic Speech Recognition Performance over Classroom Environments with Limited Data. In M. M. Rodrigo, N. Matsuda, A. I. Cristea, & V. Dimitrova (Eds.), Lecture Notes in Computer Science: *Artificial Intelligence in Education. AIED 2022*. Springer. doi:10.1007/978-3-031-11644-5_22

Schneider, B., Krajcik, J., Lavonen, J., & Salmela-Aro, K. (2020). *Learning Science: The Value of Crafting Engagement in Science Environments*. Yale University Press.

Schneider, B., Krajcik, J., Lavonen, J., Salmela-Aro, K., Broda, M., Spicer, J., Bruner, J., Moeller, J., Linnansaari, J., Juuti, K., & Viljaranta, J. (2016). Investigating optimal learning moments in US and Finnish science classes. *Journal of Research in Science Teaching*, *53*(3), 400–421. doi:10.1002/tea.21306

Schneider, B., Krajcik, J., Lavonen, J., Salmela-Aro, K., Klager, C., Bradford, L., Chen, I.-C., Baker, Q., Touitou, I., Peek-Brown, D., Dezendorf, R. M., Maestrales, S., & Bartz, K. (2022). Improving science achievement—Is it possible? Evaluating the efficacy of a high school chemistry and physics project-based learning intervention. *Educational Researcher*, *51*(2), 109–121. doi:10.3102/0013189X211067742

Schunk, D. H. (2012). *Learning theories an educational perspective* (6th ed.). Pearson.

Schwab, K. (2016). *The fourth Industrial Revolution*. World Economic Forum. https://luminariaz.files.wordpress.com/2017/11/the-fourth-industrial-revolution-2016-21.pdf

Schwonke, R., Renkl, A., Salden, R., & Aleven, V. (2011). Effects of different ratios of worked solution steps and problem-solving opportunities on cognitive load and learning outcomes. *Computers in Human Behavior*, *27*(1), 58–62. doi:10.1016/j.chb.2010.03.037

Scott, P. H., Mortimer, E. F., & Aguiar, D. G. (2006). The tension between authoritative and dialogic discourse: A fundamental characteristic of meaning making interactions in high school science lessons. *Science Education*, *90*(3), 605–631. doi:10.1002ce.20131

Sedlacek, M., & Sedova, K. (2017). How many are talking? The role of collectivity in dialogic teaching. *International Journal of Educational Research*, *85*, 99–108. doi:10.1016/j.ijer.2017.07.001

Sedova, K. (2017). A case study of a transition to dialogic teaching as a process of gradual change. *Teaching and Teacher Education*, *67*, 278–290. doi:10.1016/j.tate.2017.06.018

Seufert, S., Guggemos, J., & Sailer, M. (2021). Technology-related knowledge, skills, and attitudes of pre-and in-service teachers: The current situation and emerging trends. *Computers in Human Behavior*, *115*, 106552. doi:10.1016/j.chb.2020.106552 PMID:32921901

Sherin, M. G., & Dyer, E. B. (2017b). *Teacher self-captured video: Learning to see*. SAGE. https://journals.sagepub.com/doi/abs/10.1177/0031721717702632

Sherin, M. G., & van Es, E. A. (2007). *Using video to document changes in teachers' professional vision*. Paper presented at the Annual meeting of the American Educational Research Association, Chicago, IL.

Sherin, M. G. (2004). New perspectives on the role of video in teacher education. In J. E. Brophy (Ed.), *Using video in teacher education* (pp. 1–27). JAI.

Sherin, M. G., & Dyer, E. B. (2017a). Mathematics teachers' self-captured video and opportunities for learning. *Journal of Mathematics Teacher Education, 20*(5), 477–495. doi:10.100710857-017-9383-1

Sherin, M. G., Jacobs, V., & Philipp, R. (2011). *Mathematics teacher noticing: Seeing through teachers' eyes*. Routledge. doi:10.4324/9780203832714

Shin, N., Choi, S.-Y., Stevens, S. Y., & Krajcik, J. S. (2019). The impact of using coherent curriculum on students' understanding of core ideas in chemistry. *International Journal of Science and Mathematics Education, 17*(2), 295–315. doi:10.100710763-017-9861-z

Shuja, A. (2019). Effect of E-learning on Students' Academic Performance Mediated by Facilitation Discourse and Flexibility. *Knowledge Management & E-Learning, 11*(2), 158–200.

Silén, C., Wirell, S., Kvist, J., Nylander, E., & Smedby, Ö. (2008). Advanced 3D visualisation in student-centred medical education. *Medical Teacher, 30*(5), e115–e124. doi:10.1080/01421590801932228 PMID:18576181

Sinclair, J., & Coulthard, R. M. (1975). *Towards an analysis of discourse*. Oxford University Press.

Smith, A., Hielsher, S., Dickel, S., Soderberg, J., & Oost, E. (2013). Grassroots Digital Fabrication and Makerspaces: Reconfiguring, Relocating and Recalibrating Innovation? (University of Sussex, SPRU Working Paper). https://papers.ssrn.com/sol3/papers.cfm?abstract_id=2731835

Sohmer, R., Michaels, S., O'Connor, M. C., & Resnick, L. (2009). Guided construction of knowledge in the classroom: The troika of talk, tasks and tools. In B. Schwarz, T. Dreyfus, & R. Hershkowitz (Eds.), *Transformation of knowledge through classroom interaction* (pp. 105–129). Routledge.

Song, Y., Lei, S., Hao, T., Lan, Z., & Ding, Y. (2021). Automatic classification of semantic content of classroom dialogue. *Journal of Educational Computing Research, 59*(3), 496–521. doi:10.1177/0735633120968554

Soter, A. O., Wilkinson, I. A. G., Murphy, P. K., Rudge, L., Reninger, K., & Edwards, M. (2008). What the discourse tells us: Talk and indicators of high-level comprehension. *International Journal of Educational Research, 47*(6), 372–391. doi:10.1016/j.ijer.2009.01.001

Spronken-Smith, R. (2008). *Experiencing the Process of Knowledge Creation: The Nature and Use of Inquiry-Based Learning in Higher Education.* Semantic Scholar.

Stanford Design School. (2014). *Hasso Plattner Model Stanford design school.* Stanford Design school; Stanford Design School. https://dschool.stanford.edu/resources

Stephan, M., & Akyuz, D. (2012). A proposed instructional theory for integer addition and subtraction. *Journal for Research in Mathematics Education, 43*(4), 428–464. doi:10.5951/jresematheduc.43.4.0428

Stewart, W. H., Baek, Y., Kwid, G., & Taylor, K. (2021). Exploring Factors That Influence Computational Thinking Skills in Elementary Students' Collaborative Robotics. *Journal of Educational Computing Research, 59*(6), 1208–1239. doi:10.1177/0735633121992479

Stohlmann, M. S., Moore, T. J., & Cramer, K. (2013). Preservice elementary teachers' mathematical content knowledge from an integrated STEM modelling activity. *Journal of Mathematical Modelling and Application, 1*(8), 18–31.

Suryanti, S., Sutaji, D., Arifani, Y., Muyasaroh, M., & Zamzamy, M. (2020). Improved learning accessibility and professionalism of teachers in remote areas through mentoring development of teaching materials based on Augmented Reality. [Research Dissemination for Community Development]. *Kontribusia, 3*(1), 224–232. doi:10.30587/kontribusia.v3i1.1032

Tabatabai, S. (2020). COVID-19 impact and virtual medical education. *Journal of Advances in Medical Education & Professionalism, 8*(3), 140–143. PMID:32802908

Tal, R. T., & Miedijensky, S. (2005). A model of alternative embedded assessment in a pull-out enrichment program for the gifted. *Gifted Education International, 20*(2), 166–186. doi:10.1177/026142940502000208

Tang, K. S., Cheng, D. L., Mi, E., & Greenberg, P. B. (2020). Augmented reality in medical education: A systematic review. *Canadian Medical Education Journal, 11*(1), e81. PMID:32215146

Tang, X., Yin, Y., Lin, Q., Hadad, R., & Zhai, X. (2020). Assessing computational thinking: A systematic review of empirical studies. *Computers & Education, 148*, 103798. doi:10.1016/j.compedu.2019.103798

Tan, L. S., & Ang, K. C. (2013). Preservice secondary school teachers' knowledge in the mathematical modeling case study. In *Teaching Mathematical Modelling: Connecting to Research and Practice* (pp. 373–383). Springer. doi:10.1007/978-94-007-6540-5_31

Tan, S. Y., Al-Jumeily, D., Mustafina, J., Hussain, A., Broderick, A., & Forsyth, H. (2018). *Rethinking Our Education to Face the New Industry Era. EDULEARN18 Conference*, Palma, Mallorca, Spain. 10.21125/edulearn.2018.1564

Tekkumru-Kisa, M., & Stein, M. K. (2015). Learning to see teaching in new ways. *American Educational Research Journal, 52*(1), 105–136. doi:10.3102/0002831214549452

Ten Cate, O., & Billett, S. (2014). Competency-based medical education: Origins, perspectives and potentialities. *Medical Education, 48*(3), 325–332. doi:10.1111/medu.12355 PMID:24528467

Testoni, A. (2019). *Metodi e contenuti per un corso di chimica nel primo biennio della scuola secondaria di secondo grado. Chimica: Cosa insegnare ai vari livelli scolastici, metodologie didattiche (anche non formali), approccio integrato all'insegnamento*. Società Chimica Italiana.

The Council of State Science Supervisors. (n.d). Vision for Science Education - A Framework for K-12 Science Education: Practices, Crosscutting Concepts, and Core Ideas. COSSS. https://cosss.wildapricot.org/resources/Pictures/Overview/FRPracticesAware1hr.pdf

Thomas, J. W. (2000). *A review of research on project-based learning*. Autodesk Foundation.

Timms, M., Moyle, K., Weldon, P., & Mitchell, P. (2018). Challenges in STEM learning in Australian schools. *Policy Insights Issue 7*. Camberwell, VIC: ACER. https://research.acer.edu.au/policyinsights/7/

Tran, C., Smith, B., & Buschkuehl, M. (2017). Support for mathematical thinking through embodied cognition: Nondigital and digital approaches. *Cognitive Research: Principles and Implications*, *2*(1), 16. doi:10.118641235-017-0053-8 PMID:28280771

Tripp, T. R., & Rich, P. J. (2012a). The influence of video analysis on the process of teacher change. *Teaching and Teacher Education*, *28*(5), 728–739. doi:10.1016/j.tate.2012.01.011

Tripp, T. R., & Rich, P. J. (2012b). Using video to analyze one's own teaching. *British Journal of Educational Technology*, *43*(4), 678–704. doi:10.1111/j.1467-8535.2011.01234.x

Tsakeni, M. (2021). Preservice teachers' use of computational thinking to facilitate inquiry-based practical work in multiple-deprived classrooms. *EURASIA Journal of Mathematics, Science and Technology Education, 17*(1). https://doi.org/ doi:10.29333/ejmste/9574

Tschimmel, K. (2012). *Design Thinking as an effective Toolkit for Innovation*. In: *Proceedings of the XXIII ISPIM Conference: Action for Innovation: Innovating from Experience*. Barcelona. ISBN 978-952-265-243-0.

Tucker-Raymond, E., Cassidy, M., & Puttick, G. (2021). Science teachers can teach computational thinking through distributed expertise. *Computers & Education*, *173*, 104284. doi:10.1016/j.compedu.2021.104284

Tümay, H. (2016). Reconsidering learning difficulties and misconceptions in chemistry: Emergence in chemistry and its implications for chemical education. *Chemistry Education Research and Practice*, *17*(2), 229–245. doi:10.1039/C6RP00008H

United Nations Educational, Scientific and Cultural Organization (UNESCO). (2015). *Asia-Pacific Education Research Institutes Network (ERI-Net) regional study on transversal competencies in education policy and practice (Phase I): Regional synthesis report.* UNESCO. https://unesdoc.unesco.org/ark:/48223/pf0000231907

Uttal, D. H., Amaya, M., del Rosario Maita, M., Hand, L. L., Cohen, C. A., O'Doherty, K., & DeLoache, J. S. (2013). It works both ways: Transfer difficulties between manipulatives and written subtraction solutions. *Child Development Research*, *2013*, 216–367. doi:10.1155/2013/216367

Vamvakoussi, X., & Vosniadou, S. (2012). Bridging the gap between the dense and the discrete: The number line and the "rubber line" bridging analogy. *Mathematical Thinking and Learning*, *14*(4), 265–284. doi:10.1080/10986065.2012.717378

van der Niet, A. G., & Bleakley, A. (2021). Where medical education meets artificial intelligence: 'Does technology care?'. *Medical Education*, *55*(1), 30–36. doi:10.1111/medu.14131 PMID:32078175 van Es, E. A., Stockero, S. L., Sherin, M. G., Van Zoest, L. R., & Dyer, E. B. (2015). Making the most of teacher self-captured video. *Mathematics Teacher Educator, 4*(1), 6-19. https://www.jstor.org/stable/10.5951/mathteaceduc.4.1.0006

van Es, E. A., & Sherin, M. G. (2008). Mathematics teachers' "learning to notice" in the context of a video club. *Teaching and Teacher Education, 24*(2), 244–276. doi:10.1016/j.tate.2006.11.005

Viglialoro, R. M., Condino, S., Turini, G., Carbone, M., Ferrari, V., & Gesi, M. (2021). Augmented Reality, Mixed Reality, and Hybrid Approach in Healthcare Simulation: A Systematic Review. *Applied Sciences (Basel, Switzerland), 11*(5), 2338. doi:10.3390/app11052338

Viiri, J., & Helaakoski, J. (2014). Content and content structure of physics lessons and students' learning gains: Comparing Finland, Germany and Switzerland. In H. E. Fischer, P. Labudde, K. Neumann, & J. Viiri (Eds.), Quality of instruction in physics: Comparing Finland, Germany and Switzerland (pp. 93–110). Waxmann Verlag.

Virtanen, A., Kanerva, J., Ilo, R., Luoma, J., Luotolahti, J., Salakoski, T., Ginter, F., & Pyysalo, S. (2019). Multilingual is not enough: BERT for Finnish. arXiv preprint arXiv:1912.07076

Von Glasersfeld, E. (1989). Cognition, construction of knowledge, and teaching. *Synthese, 80*(1), 121–140. doi:10.1007/BF00869951

Voštinár, P., Horváthová, D., Mitter, M., & Bako, M. (2021). The look at the various uses of VR. *Open Computer Science, 11*(1), 241–250. doi:10.1515/comp-2020-0123

Vuorikari, R., & Punie, Y. Yves, Carreton, S., Van den Brande, L. (2016). *DigComp 2.0: The Digital Competence Framework for Citizens.* European Commission. https://publications.jrc.ec.europa.eu/repository/bitstream/JRC101254/jrc101254_digcomp%202.0%20the%20digital%20competence%20framework%20for%20citizens.%20update%20phase%201.pdf

Vuorikari, R., Ferrari, A., & Punie, Y. (2019). Makerspaces for Education and Training: Exploring future implications for Europe. *Ideas.* https://ideas.repec.org/p/ipt/iptwpa/jrc117481.html

Vygotsky, L. S. (1978). Zone of proximal development: A new approach. *Mind in society: The development of higher psychological processes*, 84-91.

Vygotsky, L. S. (1978). *Mind in society: The development of higher psychological processes.* Harvard University Press.

Vygotsky, L. S., & Cole, M. (1978). *Mind in society: Development of higher psychological processes.* Harvard university press.

Walshe, G. (2020). Radical Constructivism—von Glasersfeld. In B. Akpan & T. J. Kennedy (Eds.), *Science Education in Theory and Practice. Springer Texts in Education.* Springer. doi:10.1007/978-3-030-43620-9_24

Wasner, M., Nuerk, H.-C., Martignon, L., Roesch, S., & Moeller, K. (2016). Finger gnosis predicts a unique but small part of the variance in initial arithmetic performance. *Journal of Experimental Child Psychology, 146*, 1–16. doi:10.1016/j.jecp.2016.01.006 PMID:26895483

Weese, J. L., & Feldhausen, R. (2017). *STEM Outreach: Assessing Computational Thinking and Problem-Solving*. 124th American Society for Engineering Education Annual Conference and Exposition (ASEE 2017), Columbus, OH, USA. https://peer.asee.org/stem-outreach-assessing-computational-thinking-and-problem-solving

Weinberg, A., Dresen, J., & Slater, T. (2016). Students' understanding of algebraic notation: A semiotic systems perspective. *The Journal of Mathematical Behavior*, *43*, 70–88. doi:10.1016/j.jmathb.2016.06.001

Weintrop, D., Beheshti, E., Horn, M., Orton, K., Jona, K., Trouille, L., & Wilensky, U. (2016). Defining computational thinking for mathematics and science classrooms. *Journal of Science Education and Technology*, *25*(1), 127–147. doi:10.100710956-015-9581-5

Weintrop, D., Rutstein, D. W., Bienkowski, M., & McGee, S. (2021). Assessing computational thinking: An overview of the field. *Computer Science Education*, *31*(2), 113–116. doi:10.1080/08993408.2021.1918380

Welie, J. V. (2004). Is dentistry a profession? Part 3. Future challenges. *Journal - Canadian Dental Association*, *70*(10), 675–678. PMID:15530264

Wilson, I., & Shankar, P. R. (2021). The COVID-19 pandemic and undergraduate medical student teaching/learning and assessment. *MedEdPublish*, *10*(1), 10. doi:10.15694/mep.2021.000044.1

Windschitl, M., Thompson, J., Braaten, M., & Stroupe, D. (2012). Proposing a core set of instructional practices and tools for teachers of science. *Science Education*, *96*(5), 878–903. doi:10.1002ce.21027

Wing, J. (2006). Computational thinking. *Communications of the ACM*, *49*(3), 33–35. doi:10.1145/1118178.1118215

World Economic Forum. (2016). *What is the fourth industrial revolution?* WEF. https://www.weforum.org/agenda/2016/01/what-is-the-fourth-industrial-revolution/

World Economic Forum. (2020). *Mapping TradeTech: Trade in the Fourth Industrial Revolution*. WEF. https://www3.weforum.org/docs/WEF_Mapping_TradeTech_2020.pdf

World Health Organisation. (2021) International Data Online Updates. WHO. https://www.who.int/data)

Wright, T., Barnett, M., Hodge, B., & O'Connell, R. (2014). Investigating the effectiveness of online versus face-to-face chemistry instruction using virtual laboratory simulations. *Journal of Science Education and Technology*, *23*(6), 862–870.

Wrigley, C., & Straker, K. (2017). Design Thinking pedagogy: The Educational Design Ladder. *Innovations in Education and Teaching International*, *54*(4), 374–385. doi:10.1080/14703297.2015.1108214

Yadav, A., Krist, C., Good, J., & Caeli, E. N. (2018). Computational thinking in elementary classrooms: Measuring teacher understanding of computational ideas for teaching science. *Computer Science Education*, *28*(4), 371–400. doi:10.1080/08993408.2018.1560550

Yigitcanlar, T., Butler, L., Windle, E., Desouza, K. C., Mehmood, R., & Corchado, J. M. (2020). Can building "artificially intelligent cities" safeguard humanity from natural disasters, pandemics, and other catastrophes? An urban scholar's perspective. *Sensors (Basel)*, *20*(10), 2988. doi:10.339020102988 PMID:32466175

Yin, Y., Hadad, R., Tang, X., & Lin, Q. (2020). Improving and assessing computational thinking in maker activities: The integration with physics and engineering learning. *Journal of Science Education and Technology*, *29*(2), 189–214. doi:10.100710956-019-09794-8

Zacharia, Z. C., & Olympiou, G. (2011). Physical versus virtual manipulative experimentation in physics learning. *Learning and Instruction*, *21*(3), 317–331. doi:10.1016/j.learninstruc.2010.03.001

Zawacki-richter, O., Marín, V. I., Bond, M., & Gouverneur, F. (2019). Systematic Review of Research on Artificial Intelligence Applications in Higher Education –Where Are the Educators? *International Journal of Educational Technology in Higher Education*, *16*(39), 1–27. doi:10.118641239-019-0171-0

Zeichner, O., & Zilka, G. (2016). Feelings of challenge and treat among pre-service teachers studying in different learning environments-virtual vs. blended courses. *Journal of Educational Technology*, *13*(1), 7–19.

Zhao, Y., & Wang, L. (2022). A case study of student development across project-based learning units in middle school chemistry. *Disciplinary and Interdisciplinary Science Education Research*, *4*(1), 1–20.

Zhong, B., Wang, Q., Chen, J., & Li, Y. (2016). An exploration of three-dimensional integrated assessment for computational thinking. *Journal of Educational Computing Research*, *53*(4), 562–590. doi:10.1177/0735633115608444

Zulkifli, A. F. (2019). Student-centered approach and alternative assessments to improve students' learning domains during health education sessions. *Biomedical Human Kinetics*, *11*(1), 80–86. doi:10.2478/bhk-2019-0010

About the Contributors

Garima Bansal is an educational researcher. Her research interests lay in formative assessment, pedagogic and curricular processes in science and mathematics classrooms. She has been contributing in school and university education, and has also served development sector in education.

Umesh Ramnarain research focus is mainly inquiry-based learning. His research is on inquiry teaching and learning, and its uptake in South African classrooms characterized by diversity and complexity in terms of intrinsic and extrinsic or environmental factors. The knowledge base he has built has important implications at both the national and international levels, especially in terms of inquiry teaching in underprivileged schools.

***

Roberto Araya is a Full Professor at the Institute of Education, University of Chile. His research focuses on STEM education. He is a Ph.D. in Electrical Engineering from the University of California at Los Angeles.

Daniela Caballero is a postdoctoral fellow of the Faculty of Engineering, McMaster University. In that position, she conducts research on evaluating the impact of educational innovations on students' academic performance and experience. She has also collaborated with other educational institutions such as the Ministry of Education of Chile, and the Centre for Advanced Research in Education (CIAE) of the University of Chile.

Kennedy Chan is an associate professor in the Faculty of Education at the University of Hong Kong (HKU). He is currently a science education researcher and teacher educator with a strong commitment to understanding and developing science teachers' expertise. His research interests include science teacher professional knowledge and the use of video in teacher education. Internationally, he is a scholar in the field of pedagogical content knowledge (PCK) and teacher noticing. He has been an invited participant of the second PCK summit in the Netherlands. His research related to use of video and teacher expertise has been published in prestigious journals, including International Journal of Science and Mathematics Education, Studies in Science Education, Journal of Research in Science Teaching, Journal of the Learning Sciences and Computers and Education. He is currently an associate editor of Research in Science Education and an editorial board member of International Journal of Science and Mathematics Education and Research in Science and Technological Education.

Raija Hämäläinen works in the field of technology-enhanced learning at the Center for Research for Learning and Teaching at the University of Jyväskylä, Finland. Hämäläinen's research interests include collaboration and creativity at the technology-enhanced learning settings, workplace-learning, and teacher-student interaction. She is a well-recognised keynote speaker on novel methods (e.g. ; 7th International VET Congress 2022; International conference on research in VET 2021), an associate editor of Educational Research Review, a profiling area leader in JYU, and WP leader in a Centre of Excellence (InterLearn). Hämäläinen has wide international networks (member of Earli EC, Community Committee CSCL, Earli Sig 14 coordinator).

Catherine Hayes is Professor of Health Professions Pedagogy and Scholarship at the University of Sunderland, UK. She is a UK National Teaching Fellow and Principal Fellow of the UK Higher Education Academy. As a graduate of Podiatric Medicine in 1992, Catherine was a Founding Fellow of the Faculty of Podiatric Medicine at the Royal College of Physicians and Surgeons (Glasgow) in 2012 and was awarded Fellowship of the Royal College of Podiatry (London) in 2010. She is currently Programme Leader of the University of Sunderland's Professional Doctorate pathways for the DBA, EdD, DPM and DProf.

Abelino Jimenez received his Ph.D. degree from Carnegie Mellon University in 2019, working under the supervision of Professor Bhiksha Raj. His main research interests are machine learning, privacy preserving technologies and applied artificial intelligence in education. Abelino is a mathematical engineer from Universidad de Chile and he received his master's degree in Electrical and Computer Engineering from CMU in 2015.

Rekha Koul is an Associate Professor and Discipline Lead at STEM Research Group, School of Education, Curtin University, Australia. Dr. Koul has over three decades of teaching and research experience. She sits on various advisory and editorial boards. Her expertise lies in the development, refinement and validation of questionnaires; investigations of the effects of classroom environments on student outcomes; evaluation of educational programs; teacher-action research aimed at improving their environments and evaluation of curriculum. Her publication record includes two books authored, seven books edited, eight book chapters and many journal articles published in peer-reviewed journals. She has won research grants worth 1.2 million dollars, has delivered invited keynote addresses in various countries and also conducts teacher professional learning workshops. Rekha convenes the International Conference on Science, Mathematics and Technology Education (icSTEM), a biannual conference in different parts of the world and has established a Learning Environment Research Centre in Universitas Negeri Jakarta, Indonesia.

Joseph S. Krajcik is a University Distinguished Professor and the Lappan-Phillips Professor of Science Education. He is internationally known for his research on improving teaching and learning of science by designing, developing, implementing and testing innovative materials and environments that match evidence based research about how students learn. He is the director of CREATE for STEM, a joint institute between the Colleges of Natural Science, Education, Engineering and Lyman Briggs. Krajcik received his Ph.D. in science education from the University of Iowa. He came to Michigan State as director of the CREATE for STEM Institute in 2011 after 21 years at the University of Michigan. Krajcik has authored more than 100 manuscripts; created curriculum materials, books and software; and frequently presents at international, national, and regional conferences. He also served as co-editor of the Journal of Research in Science Teaching for five years. Krajcik's work has been recognized with numerous awards throughout his career. He is a fellow of the American Association for the Advancement of Science and the American Educational Research Association. He served as president of the National Association for Research in Science Teaching and received their Distinguished Contributions to Science Education through Research Award in 2010. In 2014, he received the George G. Mallinson Award from the Michigan Science Teachers Association for lifetime achievements in the science education field.

Sami Lehesvuori has a PhD in the field of Science Pedagogy. Adjunct professor in the field of classroom interaction. Specific research focus on dialogic teaching in science.

Tingting Li holds a doctorate degree in Curriculum and Instruction with specialization in Science Education. She is currently a Ph.D. student in Educational Psychology and Educational Technology program at Michigan State University. Her research interests focus on applying Artificial Intelligence to STEM education to support all K-12 students' deep science learning by considering psychological factors (e.g., motivational factors, social and emotional learning).

Federica Lizzi is a Primary School teacher Professor (with contract) and tutor in a degree course in the Primary Formation of Science at the University of Aquila, Italy.

Emily Adah Miller is assistant professor in the Department of Mathematics, Science, and Social Studies Education. She is the PI for Supporting MLLs in Scientific Sensemaking through an Interactional Approach to Language (MLL-SIL), and Understanding Adaptations in Multiple Literacies in Project-based Learning, and co-PI for Multiple Literacies in Project-based Learning (ML-PBL). Adah Miller's research interests are examining how elementary teachers can build on their own teaching repertoires to deepen culturally and linguistically responsive and social-justice oriented pedagogies, and, in Project-based Learning contexts, how teachers use teacher-driven adaptations to strengthen opportunities for students to productively engage in science ideas, practices and literacies.

Ayodele Ogegbo is a Lecturer in the Department of Science and Technology Education (SciTechEd), University of Johannesburg, South Africa. She holds a PhD in Science Education from the University of Pretoria, South Africa. Her special interests are physics education, technology in science education, teacher development in science education, and inquiry-based science education.

Mafor Penn is a senior lecturer at the Department of Science and Technology Education, University of Johannesburg. She holds a PhD in science education and does research in technology-enhanced science education, the affordances of Virtual and Augmented Reality (VAR) in science teaching and learning, mobile learning in science education and Inquiry Based Science Education (IBSE). She has also published some of her work in reputable journals and various national and international conferences, including SAARMSTE (Southern African countries), ISTE (South Africa), ICET, ICPE, ICESS, ICCE & END (International) conferences. ORCiD: .

Danner Schlotterbeck is a specialist in electronics and communication engineering.

Rachel Sheffield is an Associate Professor in the School of Education at Curtin University, Australia and is a passionate science educator and the Chair of the prestigious Curtin Academy which supports and celebrates excellent teaching. Dr Sheffield's research interests include women in STEM through her STEMinist community. Research grants has seen her travel to India, Indonesia and Malaysia supporting pre-service teachers, teachers and primary students (steminists.weebly.com/). She is also interested in PBL and design thinking and is currently exploring environmental sustainability in education.

Steven Ka Kit Yu has been working in the education sector for several years in teaching, research, and administrative roles. He was a secondary science teacher and a part-time lecturer in the Faculty of Education, The University of Hong Kong. He received a bachelor's degree in chemistry from The Hong Kong University of Science and Technology and a master's degree in science education from The University of Hong Kong. He is interested in science education and teacher education.

Index

T

U

V

W

Recommended Reference Books

IGI Global's reference books are available in three unique pricing formats:
Print Only, E-Book Only, or Print + E-Book.

Order direct through IGI Global's Online Bookstore at **www.igi-global.com** or through your preferred provider.

ISBN: 9781799897064
EISBN: 9781799897088
© 2022; 302 pp.
List Price: US$ **215**

ISBN: 9781799889854
EISBN: 9781799889878
© 2022; 383 pp.
List Price: US$ **215**

ISBN: 9781668451953
EISBN: 9781668451960
© 2022; 300 pp.
List Price: US$ **215**

ISBN: 9781799852001
EISBN: 9781799852018
© 2022; 355 pp.
List Price: US$ **215**

ISBN: 9781799897507
EISBN: 9781799897521
© 2022; 304 pp.
List Price: US$ **215**

ISBN: 9781799868293
EISBN: 9781799868316
© 2022; 389 pp.
List Price: US$ **215**

Do you want to stay current on the latest research trends, product announcements, news, and special offers?
Join IGI Global's mailing list to receive customized recommendations, exclusive discounts, and more.
Sign up at: **www.igi-global.com/newsletters.**

Publisher of Timely, Peer-Reviewed Inclusive Research Since 1988

IGI Global
PUBLISHER of TIMELY KNOWLEDGE

www.igi-global.com Sign up at www.igi-global.com/newsletters facebook.com/igiglobal twitter.com/igiglobal

www.ingramcontent.com/pod-product-compliance
Lightning Source LLC
LaVergne TN
LVHW081331300926
849161LV00008B/1945